MAINTENANCE AND THE SAFETY ENVELOPE

MAINTENANCE AND THE SAFETY ENVELOPE

by
H. C. Howlett II

TECHSTAR
Pocatello, Idaho

MAINTENANCE AND THE SAFETY ENVELOPE

First Edition Published August, 2003.
Printed in the United States of America.

15 14 13 12 11 12 11 10 9 8 7 6 5 4 3 2

ISBN 1-57614-023-7

TECHSTAR
P.O. Box 2651
Pocatello, ID 83206-2651
(208) 234-4119

www.techstaronline.com
www.techstarbooks.com
www.techstarpublishing.com
www.techstartraining.com

Preface

Since publication of the first edition of **The Industrial Operator's Handbook** in October, 1995, we have been overwhelmed with the response of industry to the concepts presented in the book. Thousands of readers have consulted the text for application in their industries and thousands more—operators, maintenance technicians, laboratory analysts, engineers, and managers of all levels—have participated in the seminars we have presented across the United States, Canada, and other countries.

Our travels have taken us to a myriad of North American electrical generators powered by nuclear, coal, gas, and oil fuels as they provide the electrical needs of their communities. We have visited Department of Energy sites throughout the U.S. whose teams engage daily in sampling, characterizing, processing, packaging, shipping, and storing radiological contaminants and other hazardous waste. And we have evaluated the activities and taught the staffs of oil refineries and chemical production plants in places as far-ranging as Trinidad.

Nearly always, we are impressed by the knowledge, skill, and dedication of these teams and their leaders. At the same time, we are amazed at what they are able to accomplish in accident prevention as cutbacks in the sizes of workforces and resources have swept each industry. Clearly, "doing more with less" is a challenge to the best teams as they try to balance mission accomplishment with reasonable safety.

Yet, as we hear the stories and study the reports of our clients' accidents and near misses (and those of industry at large), we are now even more convinced of the need to instill disciplined habits of systematic operations and maintenance to *all* elements of the workforce—for, in nearly every case we review, the events could have been easily prevented through application of a few of the simple techniques discussed in Section III (Vital Operating Skills) of **The industrial Operator's Handbook, 2nd Edition**.

For that reason, we have begun a new stage in our mission to provide insight and instruction to the supporting elements of industrial teams (elements such as maintenance and engineering) regarding application of the principles and skills presented in the main work. In that vein, we have set out to produce a series of supplemental resources focusing on specific topics (and dedicated to specific work groups).

This book is the first in that series. It is dedicated to the maintenance teams whose efforts provide to the operating crews equipment which functions as intended by design—machinery and processes which are safe and reliable. Without properly performed maintenance, neither objective is possible.

We hope you enjoy your journey!

Contents

Chapter 1

Introduction and Review

Accidents are almost always the cumulative result of a chain of apparently minor deviations from known standards, practices, and procedures. Viewed in isolation, no single deviation appears important. But together, they form unstable circumstances—the conditions which support failure.

In **The Industrial Operator's Handbook**, we learned that safe operation of equipment and processes relies on far more than good design and proper construction. Excellent automobiles are damaged and occupants injured when drivers fail to control their vehicles. Aircraft crash when professional pilots neglect to execute tasks in accordance with their training and in compliance with procedures which prescribe safe task performance, or when their maintenance counterparts fail to restore equipment to design specification. And industrial facilities experience fires, explosions, and uncontrolled hazardous material releases which result when operating and maintenance teams incorrectly align or improperly control machinery or processes.

We found that "accidents" are usually not accidents at all; rather, they are the logical (frequently predictable) outcomes of accumulated deviations from known standards, practices, and procedures—a repetitious collage of contributors, in most cases forming a familiar pattern or "anatomy" much like the building of an unstable house of cards.

And, as we reviewed a series of accidents, we reinforced the principle that equipment reliability and safety are achieved through conservative design, superior construction, proper operation, and precise maintenance—each phase successively contributing to a safety envelope which, if observed, offers reasonable risk minimization against exposure to credible hazards.

We discovered that, through design, the boundaries of equipment performance are established for normal circumstances and also for predictable abnormalities. In construction, the designer's concept is executed with specified materials and joined by prescribed processes. During operation, skilled and knowledgeable team members control the equipment in accordance with limits determined during design. And through maintenance, equipment is restored to current design requirements when components degrade.

Our journey taught us that the integrity of the safety envelope depends upon success in each phase. Deficient performance at any juncture may invalidate the envelope and jeopardize team members, the surrounding public, or the environment. And, as the title of the book implied, we focused on the role of the operator, emphasizing that there is no substitute for alert, well-trained operators controlling equipment within specified bands in accordance with approved procedures.

This book emphasizes the parity of maintenance teams in accomplishing the same objectives of equipment safety and reliability. Without proper *maintenance*, safety and reliability objectives cannot be achieved and a breach of the protective envelope is inevitable. Therefore, if operating equipment is to remain reliable, precise equipment maintenance must supplement conservative design, superior construction, and proper operation. Subsequently, the theme of this book is that *there is no substitute for alert, well-trained maintenance technicians maintaining equipment within specified bands in accordance with approved procedures!*

The Causes of Accidents

As we stated at the beginning of this chapter, accidents are almost always the cumulative result of a chain of apparently minor deviations from known standards, practices, and procedures. Viewed in isolation, no single deviation appears important. Together, however, they form unstable circumstances—the conditions which support failure.

Accident Anatomy

In **The Industrial Operator's Handbook** [Chapter1, Anatomy of an Accident], we reviewed the tragedy of Eastern Airlines Flight 401, a story of compounding errors in which the aircrew failed to monitor critical operating parameters. We found lapses in leadership, errors in communication, ignorance of warnings, failures in monitoring critical parameters, and deficiencies in emergency response. Together, these factors contributed to an accident that killed 99 passengers and crew. Yet, the only thing actually wrong with the aircraft was a couple of burned out indicator light bulbs.

In almost every accident that we study, we find that a combination of factors is necessary for the accident to occur. It is very similar to the process of building a house of cards. As cards are added to the stack, the instabilities of each contribute to the greater instability of the whole. Finally, the last card—an "initiating factor" in accident study terminology—is placed and the house falls down. And the collapse of the house is greater in consequence than for a single card. That was the case for the Three-Mile Island accident in 1979, the Chernobyl accident in 1986, the *Exxon-Valdez* accident in 1989, and the fatal gas release at Bhopal in 1984. In each accident, a combination of factors was necessary for—and exacerbated the outcome of—the final consequential event.

Common Contributing Factors

In **The Industrial Operator's Handbook** [Chapter 2, Common Components of Accidents], we identified and learned about twenty of the most common contributors to accidents—errors with which we are all familiar. Those "common components" include:

1. Defective night operations
   (usually resulting from relaxed attention, fatigue, and poor visual conditions)
2. Unfitness for duty
   (whether due to fatigue, inadequate training, drugs or alcohol, medicine, or emotional stress)
3. Incomplete communication
   (people assuming that others have the same information and mental pictures that they themselves do)
4. Inadequate turnovers
   (poor exchange of information when one work group or team member relieves another)
5. Ignorance of equipment operating characteristics
   (people who don't understand their machines and how they function)

6. Non-compliance with procedural requirements
   (not having, not understanding, or not using the procedures which have been developed to regulate operating and maintenance tasks)
7. Failure to monitor equipment status
   (not properly monitoring the parameters or values associated with equipment operation which are most closely related to failure)
8. Deficient equipment maintenance
   (not keeping equipment within design tolerances and functions)
9. Failure to independently verify critical tasks
   (not second-checking actions or functions which are infrequently performed, complex in nature, or high consequence if improperly performed)
10. Sense of invulnerability
    (having an attitude that "it can't happen to me")
11. Acceptance of abnormalities
    (letting conditions that are incorrect exist without remedy)
12. Ignorance of warning signs
    (not paying attention to the signs that almost always precede undesirable events)
13. Capitulation to production pressure
    (letting schedule pressure override critical thinking)
14. Inadequate emergency preparedness
    (not being ready to respond to abnormal events which are probable and predictable)
15. Deficient organizational staffing and structure
    (not having the right number of people with the right skills in the right places)
16. Poor leadership
    (leaders who fail to effectively coordinate and direct their teams toward task accomplishment)
17. Insufficient management oversight
    (leaders who don't consistently coach, correct, and train their teams toward improvement and high standards)
18. Faulty teamwork
    (when team members and team leaders don't coordinate with one another to pull in the same direction at the same time)
19. Degradation of operating limits
    (allowing equipment to be operated and maintained outside the boundaries envisioned and prescribed by the designers)
20. Defective design
    (equipment for which the design and construction do not support reliable operation and maintenance)

A review of these contributing factors indicates that most are the result of human operating and maintenance errors. Few are the result of poor equipment design or construction.

That certainly is no surprise to those who operate and maintain power plants, oil refineries, chemical production plants, or pipelines. Maintenance and operating teams comprise the front lines of their industries and, subsequently, have the greatest opportunity to misalign, improperly maintain, or improperly operate their equipment.

Fortunately, most of the common contributors can be stopped by disciplined and systematic operating and maintenance habits. Yet, in order to establish such habits, plant staffs must be trained in the purposes and forms of such habits. Then, once established, they must be enforced.

As we interview people who have been involved in accidents or near misses (or as we read the reports of accidents and the testimonies of people involved), the bad habits, procedural deviations, failed lockout/tagout, or poor communication techniques that contributed don't individually look very important to them. Only in retrospect are they able to see how the "little" factors combined to cause a dangerous situation. That seems to be the nature of accidents, and it is only through accident study that people become aware of how important the little things are. Yet, teaching the concept of the importance of little things is one of the greatest difficulties for sports coaches, combat team leaders, and industrial team leaders.

Practices That Prevent Accidents

Industries such as nuclear powered electrical generation and commercial aviation have done a better job of studying accidents than most others. They have learned this pattern of little "unimportant" mistakes stacking up. Therefore, they have established (and they enforce) rigorous operating and maintenance practices which seem sometimes to be too conservative. Yet, the habits are very much like checking the blind spot before changing lanes in defensive driving. Seldom do you find someone there, but the habit takes so little effort that, once established, it offers great protection. Therefore, disciplined defensive drivers find it unwise not to take the action.

12 Vital Skills

Similarly, industrial "defensive driving" habits have been learned and taught in many industries and activities for years, often beginning with high risk activities such as flight deck operations aboard aircraft carriers. In **The Industrial Operator's Handbook** [Chapter 7, Controlling Equipment and Processes], we identified twelve of these "vital operating skills". These defensive habits and practices include:

1. Conducting pre-job briefings
2. Establishing, understanding, and using procedures
3. Monitoring critical operating parameters
4. Performing independent verification of critical tasks
5. Communicating vital information (with emphasis on three-part communication)
6. Keeping logs and recording critical parameter values
7. Recognizing and responding to abnormalities in equipment and processes
8. Combating equipment emergencies and casualties
9. Overseeing maintenance, modification, and testing of equipment
10. Isolating energy hazards
11. Performing on-the-job training, and
12. Performing work group turnovers

None of these habits is unusual in industry or hard to do. But the habits have to be trained and enforced. If procedure compliance is not trained and enforced, experience indicates that people will fall into the habit of not following procedures (or following them in such a loose sequence that problems eventually occur). If hazardous energy isolation (lockout/tagout) is not trained and enforced, energy isolation problems eventually become common. If proper communication habits are not trained and enforced, poor communication eventually becomes a factor in an accident. If pre-job briefings are not conducted or are not specific and timely, then people don't think through what they're about to do before they do it. Obviously, competent and frequent oversight and coaching by good leaders is necessary to achieve such a disciplined operational state.

Application to Maintenance

The objective of this book is to remind all of us in industry that these same vital skills apply to the maintenance process as well as to operations. In that vein, we have selected four case studies in which maintenance failures played vital roles in the accidents which occurred.

The first case study emphasizes that maintenance technicians must not only be well-trained, but also alert. The second case study highlights the need for equipment to be maintained within specified bands if the integrity of the safety envelope is to be guaranteed. The third case study reminds us that approved procedures usually govern maintenance processes. And, finally, the fourth case study stresses the idea that all the protective barriers in the world cannot compensate for uncommitted or inattentive people.

So, as we proceed with our study, consider how many of the common contributors affected the accidents that we review—and how many of the accidents could have been stopped by the simple application of soft barriers—the habits of disciplined maintenance.

Chapter 2

Maintenance and the Safety Envelope

Maintenance units perform a vital function in guaranteeing the integrity of the safety envelope for industrial facilities. If they fail to keep the safety envelope armor intact, the consequences can be disastrous. In the Three-Mile Island accident (March, 1979), an operating/maintenance team failed to realign the outlet valves to the auxiliary feed water pumps after performing a functional check on the pumps. In the Piper Alpha North Sea oil platform explosions and fire (July, 1988), a maintenance team leader failed to inform the operating shift manager that a relief valve was missing from a gas compression system. In the Pasadena, TX, Phillips 66 chemical complex explosions and fire (October, 1989), a maintenance team bypassed protective features in a chemical reactor isolation leg, releasing tons of hot, pressurized flammable gases to atmosphere. In the Tampa, FL, Gannon Power Plant hydrogen explosion (April, 1999), a maintenance team member removed a generator access cover that formed part of the hydrogen coolant boundary before the hydrogen had been purged from the system.

In reviewing events such as these, it is clear that safety depends upon those who maintain equipment, as well as upon the designers, builders, and operators. The focus of this book is the importance of maintenance teams in accomplishing the objectives of equipment safety and reliability. For without proper maintenance, safety and reliability objectives cannot be achieved. If equipment is to remain reliable, precise equipment maintenance must supplement conservative design, superior construction, and proper operation. Supporting that premise, the theme of this work is that *there is no substitute for alert, well-trained maintenance technicians maintaining equipment within specified bands [the safety envelope] in accordance with approved procedures!*

Maintenance Terms and Definitions

During the life of components and systems, maintenance and/or modification will usually be required. If improperly performed, either can void the original safety analysis.

As we learned in **The Industrial Operator's Handbook** [Chapter 3, Systematic Industrial Operations], maintenance is the process of restoring equipment to current design specifications. We also made a further distinction between this restorative process known as *corrective* maintenance and *preventive* maintenance which is a process to detect and forestall degradation. Periodic examination for wear and lubricating moving parts would, for example, be a form of preventive maintenance.

Since the safety analysis for an operating complex is based upon precise equipment design, deviations from valid maintenance requirements and practices may invalidate the conclusions of the safety analysis. For example, if repair parts or materials don't meet specification, the equipment in which they are used may not function as designed. Accordingly, strict control of the maintenance process is critical to safe operations.

Modification, by definition, is a change to original design specifications. When contemplating modification to any part, component, or system affecting safety, the safety analysis must be revisited and, if necessary, redone to accommodate the change. That's why manufacturers' warranties are invalidated when unauthorized modifications are performed.

Since improperly performed maintenance and modification can adversely affect safe operation, restrictions are imposed on their planning, performance, control, and review to ensure that no new hazards are introduced and that protection for all originally analyzed hazards is still intact. The principles of control are best summarized in two simple rules: "Maintain within design specifications" and "If modifications are contemplated, evaluate how they will affect the safety analysis."

The Safety Envelope

Most equipment and processes are well-designed for their intended use. But their "safeness"is predicated upon their components remaining within design tolerances so that the equipment is reliable—functions as designed when required.

Maintenance, then, plays a vital role in preserving the integrity of the safety envelope governing the process or equipment. So, before we continue, let's review the definition, purpose, and elements of the safety envelope.

Safety Analysis

Hazards of varying degrees are an inherent part of using just about every kind of tool, equipment, or process. Whether a tool is simple or complex, we assume risk to gain advantage. Yet, we also attempt to limit associated risks to levels that are reasonable.

As we learned in **The Industrial Operator's Handbook** [Chapter 4, Boundaries of Safe Operation], achieving limited risk in an acceptable fashion starts with *safety analysis*—the process of identifying hazards, determining the risk associated with each hazard, and erecting protective barriers and boundaries of design, construction, operation, and maintenance to protect against hazards that are deemed probable and unacceptably harmful.

Protecting against hazards involves the development of safety schemes. Whether for table saws in home wood shops, automobiles, commercial aircraft, power generating facilities, or chemical production plants—all are bound by reasonable standards of design, construction, testing, operation, maintenance, and modification which are applied to reduce to an acceptable level exposure of people and property to the hazards that accompany the subject technology and its processes.

Safety Envelope

Some of the protections in safety schemes are physical in nature and some are merely mental barriers. Yet, the composite of boundaries, whether formal or informal, is often termed a *safety envelope*.

Safety envelopes—*graphic or verbal representations of the margins and boundaries of safety established for critical parameters of machinery design, construction, operation, and maintenance*—are the offspring of safety analysis. They are developed to provide reasonable safety for the operating staff, the general public, and (for processes which can affect it adversely) the environment. When properly devised and applied, they offer great protection against exposure to identified hazards.

Safety envelopes are complex because they must often consider and protect against dozens of hazards. Further, the means of protection (risk management techniques) vary as widely as the hazards. Just as in driving, sometimes protection for a certain event is derived from a hard barrier like frame design. More often, hard barriers like frame design are combined with soft barriers in the form of good operating practices (like defensive driving techniques), and then backed up with personal protective equipment and clothing (like seat belts).

Envelope Characteristics

In **The Industrial Operator's Handbook**, we learned that safety envelopes have three distinct regions: 1) the failure zone—an area of harm or consequence caused by exceeding one or more critical parameters governing equipment or process control, 2) margins and boundaries to buffer against failure, and 3) the zone of normal operations—the region in which machinery control should normally reside. [See Figure 2-1, Margins of Safety.]

The margins and boundaries of protection—sometimes physical, sometimes procedural, and sometimes administrative—are established through proper design and construction in accordance with prescribed codes, compliance with operating and maintenance limits, and through observance of best industry operating and maintenance practices. They are promulgated to the user through design docu-

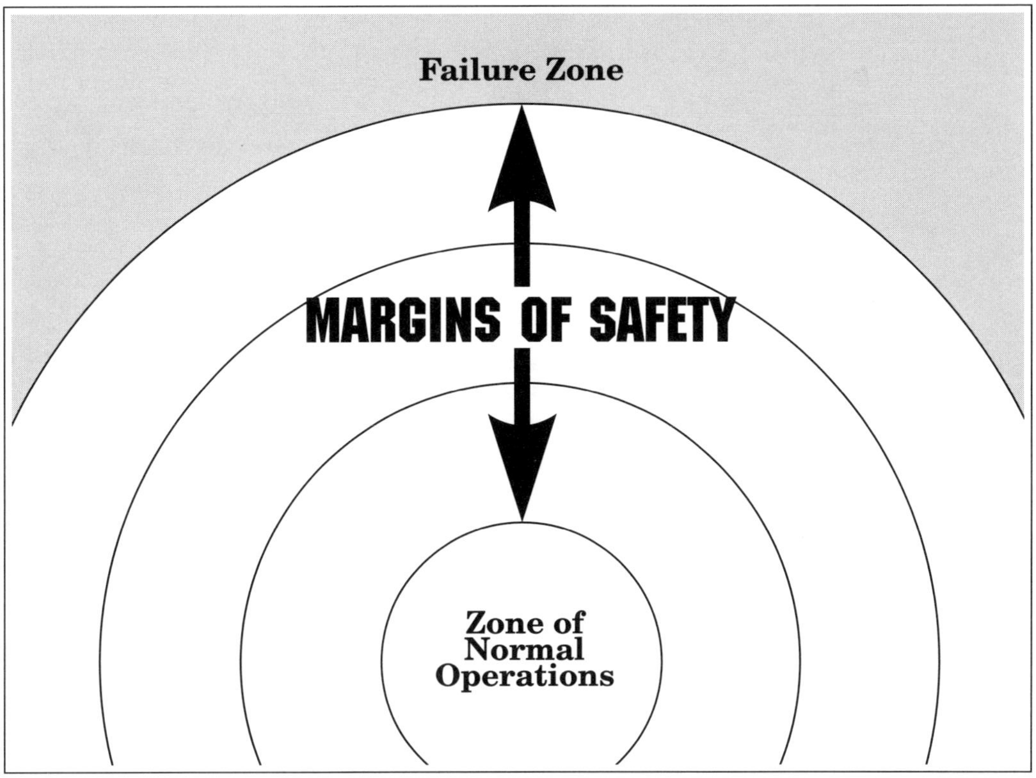

Figure 2-1: Margins of Safety; a graphical representation of the Safety Envelope. [TECHSTAR]

mentation, technical manuals for the equipment involved, and written procedure for how the equipment is to be used and maintained.

For aircraft, as an example, not only are there rigorous design and construction standards and records that must be met and maintained (regulated by the FAA), but the limits and good practices for proper operation and maintenance are provided through operations manuals, maintenance manuals, training standards, and the pilot's operating handbook. Weakness at any of those points threatens the integrity of the safety envelope.

The Human Factor

Yet, all of the excellent design, construction, built-in safeguards, precautions, and instructions are only a starting point for safe and successful operation. They are not a guarantee against damage to people, equipment and facilities, or the environment in which they are contained. They do not eliminate risk. Rather, they provide reasonable risk minimization *when affected people use them wisely.* Nevertheless, everyday, people DON'T use them in accordance with law or best industry practices.

The lesson is this: The best of equipment (well-designed and well-constructed) can, through improper operation and maintenance, result in serious injury or death. You can even be killed in a Mercedes if you drive it over a cliff. So, just as in driving or flying, the equipment itself is only a small portion of the safety envelope. The training, knowledge, commitment, coordination, and oversight of those who use and maintain the equipment are, ultimately, the larger part of safety.

Hard and Soft Barriers

Protective barriers (for hazards of concern) can range from "hard" to "soft". [See Figure 2-2.] By their nature, hard barriers are tangible and usually easy to see and assess. Typically, they physically block exposure to a hazard.

Examples of hard barriers include equipment design features, protective interlocks, guards and shields, hazardous energy isolation devices (lockouts), and personal protective clothing and equipment. A blade guard on a table saw is a clear example of a hard barrier.

Soft barriers (like good command and control within an operating crew) are far less visible because they are, for the most part, related to how well-trained, disciplined, supervised, committed, and alert are the equipment users. As such, soft barriers are usually more difficult to see and easier to bypass.

Examples of soft barriers include training, procedures, supervision, controlled communications, job briefings, independent verification, and a host of other long-proven work practices.

This combination of hard and soft barriers together forms a safety net—sometimes called a safety system, a safety envelope, or a safety case—that, when applied with discipline, *reasonably reduces* the risk of exposure to the hazards that are characteristic of the technology. It must be emphasized that safety envelopes seldom eliminate all risk. There is assumed risk in the use of almost any tool or piece of equipment, or in almost any activity. The goal, then, of the safety envelope is to *reasonably minimize* risk.

Defense In-Depth

One of the most important concepts in safety analysis is that of *defense in depth*—the idea that, for some plausible circumstances, no single barrier may be sufficient to stop a harmful event. Rather, the safety envelope derived from safety analysis is a fabric woven of many individual barriers—both hard and soft—to form a protective net. No single cord may be able to arrest the fall, but all of the cords together form effective protection for plausible events.

But the methods of providing interlocking protections for human actions are not as straightforward as they are for machines. That's why disciplined equipment operation and maintenance rely not just on excellent training (which, of course, is the foundation of intelligent control), but also on the use of written procedures (since memories fail), the use of independent verification for critical actions (since sometimes we miss something), the use of formal communication techniques including commands, repeat-backs, and corrections (because sometimes we don't hear things as they were intended), and the use of job briefings (because they cause us to focus on what we are doing right before we start doing it). In addition to these, there are many other work practice techniques that have proven useful and effec-

Figure 2-2: Barriers; a graphical representation of "hard" barriers (walls) and "soft" barriers (fences and warning tape). [TECHSTAR]

tive in limiting the number of human errors that occur because of our human frailties. [See **The Industrial Operator's Handbook, 2nd Edition**, Chapters 7-19.]

The Most Important Barrier?

It is impossible to determine whether equipment reliability (and associated hard barriers) or human reliability (and associated soft barriers) is the more important factor in the safety envelope prescription. Admiral Rickover, father of U.S. Naval Nuclear Propulsion, said it like this:

> *Over the years, many people have asked me how I run the Naval Reactors Program so that they might find some benefit for their own work. I am always chagrined at the tendency of people to expect that I have a simple, easy gimmick that makes my program function. They are disappointed when they find out there is none. Any successful program functions as an integrated whole of many factors. Trying to select one aspect as the key one will not work. Each element depends on all the other elements... [A]ll of these elements must mesh for the system to work. You cannot separate out and use the pieces which you like, and discard those which are "too hard".*

As a result, Rickover put as much effort into developing people and disciplined habits of operation and maintenance (soft barriers) as he did into developing top-notch equipment (hard barriers). The history and effectiveness of the U.S. Naval Nuclear Propulsion Program is a testament to the effectiveness of that strategy. One need only observe operations in the maneuvering room of a nuclear submarine, the control room of a U.S. nuclear power plant, flight deck operations aboard an aircraft carrier, or the cockpit of a major air carrier like Delta to see how the barriers of disciplined human performance are integrated with the hard barriers of design.

Keeping the Fabric Intact

Since safety envelopes integrate many protections, it follows that envelopes which are not kept intact may not be effective in arresting a "fall". All the cords in the net must be in place and secure if the net is to be effective. It is the netting *system* that works together to arrest the fall—an undesirable event, usually the result of human error—that might otherwise be injurious.

Because safety envelopes are a conservatively designed fabric system, there is a danger to which some industrial teams fall prey—the belief (whether passively or actively exercised) that individual cords of the safety net are not all necessary and may be selectively removed. ("See, we didn't need it. We didn't use it and nothing happened.") Such teams don't understand that the envelope is designed and constructed predominantly for *abnormal* situations—circumstances that are often triggered or compounded by the mistakes people make. The safety margin that seemed so conservative during *normal* circumstances suddenly becomes very thin during a casualty or emergency. Therefore, the removal of one or more of the cords in the net must be viewed with great concern by industrial teams and their leaders. Sadly, some teams dismantle their safety envelopes through ignorance born of poor training and supervision. Others dismantle their envelopes with impunity—such as occurred in the grounding of the *Exxon-Valdez* in Prince William Sound in March, 1989, and the Chernobyl Atomic Power Station accident in April, 1986.

In **The Industrial Operator's Handbook, 2nd Edition**, we reviewed several maintenance-related accidents that illustrate failures to maintain safety envelopes. The

1994 Utility Crew Electrocution [Chapter 8] and the 1999 TOSCO Unit 50 Oil Refinery Fire [Chapter 17] were obvious failures to observe the safety envelope. Similarly, the 1991Continental Express Accident [Chapter 19] was caused by a maintenance crew's failure to preserve and restore the aircraft's control surface integrity—an essential factor in the safe operating envelope. The 1984 Bhopal disaster [Chapter 7] involved an ongoing failure of maintenance teams to preserve and restore the facility's design features for safe operation, *and* a maintenance team's failure to properly perform a routine piping flush within the safety envelope constraints for that specific evolution.

We can see that many of the cords in the safety net are soft barriers—barriers which are based more upon the performance of people than on the initial design and construction of the equipment. Consequently, the effectiveness of a safety envelope depends primarily upon the people who use it and upon the people who maintain it. When equipment is controlled and maintained in a disciplined fashion and in accordance with current best practice guidelines (called defensive driving in the realm of automobile operation and called consensus standards in other areas of industry), well-designed and constructed equipment is reasonably safe in use.

Think About It

So, as you study and review the four maintenance-related case studies in this text, also ask yourself:

- Which Common Contributors affected the accidents?
- How were the Safety Envelopes violated?
- What "Barriers" were ignored?
- What habits of disciplined maintenance could have prevented these accidents?

Chapter 3

Even Small Errors Have Major Effects

In the last chapter, we reviewed the concept of safety envelopes and were reminded that protective barriers extend far beyond conservative design and superior construction. Equipment must also be operated and maintained in accordance with the designer's specifications if it is to remain reliable—to function as designed when required.

In this chapter, we will focus on the importance of alert, well-trained maintenance technicians as we review a "near miss" event that resulted from a simple error committed by two aircraft mechanics. The clear lesson is this: The success of maintenance (and the subsequent validity of the safety envelope) ultimately depends upon the training, commitment, and alertness of technicians and their leaders who are charged with repairing equipment.

Eastern Airlines Flight 855, May 1983

On May 5, 1983, a Lockheed L-1011 [Figure 3-1] operated by Eastern Airlines as Flight 855 experienced a loss of lubricating oil to all three engines while en route from Miami International Airport to Nassau, Bahamas. As the flight crew experienced engine problems, they elected to return to Miami rather than attempt a

Figure 3-1: Typical L-1011 Aircraft. [NASA Photo]

landing at their port of destination. The flight crew was forced to shut down all three engines sequentially. They were able to restart the No.2 engine in time to complete their approach into Miami. The loss of lubricating oil was determined to have been caused by the installation of chip detectors without the necessary O-ring seals during routine maintenance the previous evening.

A Routine Flight

Eastern Flight 855 was a regularly scheduled passenger flight from Miami, Florida, to Nassau, New Providence Island, Bahamas—a distance of about 160 miles. The aircraft, a Lockheed L-1011, had been in service with Eastern Airlines since November of 1976, and was powered by three Rolls Royce Model RB-211-22B engines (No. 1, No. 2, and No. 3) which had been restored and installed on April 11, 1983, November 22, 1982, and December 29, 1982, respectively.

The flight crewmembers were both experienced and well-rested on the morning of May 5, 1983. The captain (pilot) had 12,045 hours of flight time and had been off duty for more than 24 hours before the flight. The first officer (co-pilot) was also a supervisory check captain with 16,946 flying hours and had been off duty for 22 hours before the flight. The second officer (flight engineer) had flown 9,027 hours and had been off duty for 24 hours before the flight.

When the aircraft left Miami at 8:56 A.M., it was expected to be nothing more than a routine 37 minute flight. All pre-flight checks and checklists were conducted normally; the flight engineer's "walk-around" inspection disclosed no indications of oil leaks, and there were no abnormal engine instrument readings during engine start or the taxi-to-takeoff. Weather conditions at Nassau were 500 feet scattered clouds, 1,000 feet broken, 7,000 feet overcast, with visibility 4 miles in light rain and fog. Miami International Airport had a 2,300 foot ceiling with 7 miles visibility, and no significant weather.

Takeoff and Climb

Flight 855 departed Miami International's Runway 27R at 8:56 A.M. with 162 passengers and 10 crewmembers on board. The flight was on an instrument flight rules (IFR) plan to Nassau because of weather conditions with an estimated flight time of 37 minutes.

Takeoff and initial climb to Flight Level (FL) 230 (23,000 feet) were uneventful. The flight engineer monitored the activities of the captain during takeoff and scanned his own panel for abnormal indications—finding all engine instrument readings to be normal. He also checked the electrical system and engine instruments, called Eastern Airlines at Miami to report their departure time, completed the climb checklist, called the Eastern Airlines Miami Technical Center to establish contact with the flight dispatcher, and continued to scan the engine instruments.

From takeoff until reaching level flight at 23,000 feet at 9:08 A.M., the oil quantity gauges did not fluctuate, and all three indicators showed about 18 quarts. ("Full" indication is 21 quarts.)

At 9:10 A.M., the flight was cleared to descend to 9,000 feet, and the captain began the descent. At 9:11 A.M., the flight engineer contacted Nassau approach control to report their descent through FL 200 (20,000 feet), about 70 miles from Nassau. Nassau reported the weather as ceiling 1,000 feet and visibility 5 miles, with thunderstorms to the north, and 500–800 foot ceilings to the east and south. Nassau approach also informed Flight 855 that it was following about 30 miles behind a light twin-engine aircraft. The flight engineer then prepared the landing card data and read the descent and in-range checklist in preparation for landing at Nassau. While preoccupied with preparation of the landing data, he did not scan his panel or the oil quantity and pressure gauges for about 5 minutes.

First Problem

As Flight 855 descended through 15,000 feet, the captain informed the flight engineer that he had a low oil pressure warning light for the No. 2 engine (center engine in the tail). The flight engineer had just finished the landing card data, and, after the captain's communication, observed the Number 1 and 3 engine instruments both indicated oil quantity at about 15 quarts with oil pressure in the "green" (acceptable) range. The No. 2 engine, however, indicated oil quantity as 8 quarts, with pressure fluctuating between 15 psi and 25 psi. The flight engineer was not concerned with the 15 quart quantity indications for engines No. 1 and No. 3, because he expected to see fluctuations of plus or minus 3 quarts.

No. 2 Engine Shutdown Ordered

The flight engineer then advised the captain of the observed instrument readings, adding that the minimum oil pressure required for normal operation was 30 psi. The captain ordered a shut down of engine No. 2 and started the auxiliary power unit (APU). This was the standard procedure in response to this kind of emergency.

At 9:15 A.M. while at 12,300 feet and now about 50 miles from Nassau, the captain requested clearance back to Miami. The first officer concurred with the captain's election to return to Miami rather than continue their descent to land at Nassau. This decision was based on the deteriorating weather conditions reported at Nassau (and also observed on Flight 855's radar) as well as anticipated delays in landing because of Nassau's "non-radar" environment and air traffic (the twin-engine aircraft) which would land ahead of them—they believed that they could land at Miami sooner than they could at Nassau.

So, at 9:15 A.M., Flight 855 was cleared to make a 180° turn and maintain altitude at 12,000 feet. At 9:18 A.M. the flight was cleared to climb to FL 200 (20,000 feet). During this time, the flight engineer completed the engine shutdown checklist and the secondary items on the emergency checklist. These tasks required about 4 minutes to complete, and he stated that he was not constantly monitoring the indications for engines No. 1 and No. 3 during this time.

Additional Failures

As the aircraft climbed through 15,000 feet, the low oil pressure light for the No. 3 engine illuminated, followed almost immediately by illumination of the No. 1 engine low oil pressure light—and the oil quantity gauges for all 3 engines now were reading zero. At 9:23 A.M. the flight leveled at 16,000 feet and they informed Miami Air Route Traffic Control Center (ARTCC) of the engine gauge readings, but stated that "we believe it to be faulty indications since the chance of all three engines having zero oil pressure and zero quantity is almost nil". The flight engineer contacted the Eastern Airlines maintenance personnel (Miami Technical Center) to determine whether there was a common electrical source that could affect the instruments for all three engines. The captain stated that he assumed the problem of low oil pressure and zero quantity for the three engines was a faulty indicator problem since the likelihood of simultaneous oil exhaustion in all three engines was "one in millions I would think".

Miami Technical Center called back to Flight 855 to inform them that the Number 2 AC bus was the common power source for the oil quantity instruments. The Flight Engineer checked the appropriate circuit breaker, but found no discrepancies.

At 9:28 A.M., the No. 3 engine failed. The aircraft, now about 80 miles from Miami, began a gradual descent—powered only by the No. 1 engine. The flight crew now realized that the indications of zero oil pressure and quantity were correct, and not merely a gauge problem. At 9:29 A.M., Miami ARTCC cleared Flight 855 to descend to any altitude required and to fly directly to Miami International Airport for a landing on Runway 27L. The flight engineer called the senior flight attendant to the cockpit and instructed her to prepare the cabin for ditching. The senior flight attendant then briefed the other flight attendants, who then began the necessary pre-ditching preparations

The flight crew was unsuccessfully attempting to restart the No. 2 engine when, at 9:33 A.M., the No. 1 engine failed. The airplane, now at a 12,000 foot altitude, was "flying"—or rather, "gliding"—without any engines operating about 55 miles from Miami. As a result, the rate of descent increased to about 1,600 feet per minute. It was descending at 600 feet per minute with one engine operating.

At 9:33 A.M., Flight 855 advised Miami ARTCC that no engines were operating; and shortly thereafter, the flight engineer announced over the public address system that "ditching is imminent". Meanwhile, the Coast Guard and other potential rescue vessels were notified of the emergency. All aircraft and surface vessels were alerted and either underway or preparing to get underway by 9:44 A.M.

The captain unsuccessfully tried a second time to restart the No. 2 engine, and then (also unsuccessfully) attempted to restart the No. 1 and No. 3 engines. Simultaneously, the flightcrew was proceeding through the ditching checklist.

A third attempt to restart the No. 2 engine was finally successful at 9:38 A.M.—with the airplane at 4,000 feet and about 22 miles from Miami. The descent was finally arrested at 3,000 feet; and, at 9:40 A.M. they began a gradual climb to 3,900 feet. At 9:43 A.M., Flight 855 leveled at 3,900 feet and began the final descent for landing at Miami International Airport. They landed at 9:46 A.M.

Engine Fire

After Flight 855 landed, Air Traffic Control reported to the captain that smoke was coming from the No. 1 engine. The captain discharged the fire bottles in the engine and the smoke stopped. He later discharged the fire bottles in the No. 3 engine after smoke was reported to him by ground personnel. The airport crash/fire/rescue (CFR) unit used 300 gallons of water and 20 gallons of fire-suppressant foam to extinguish the fire. The aircraft was towed to the gate because it could not taxi with power from only one operating engine (No. 2), and the passengers deplaned normally at the terminal.

Fortunately, there were no injuries or fatalities as a result of this occurrence. However, it should be noted that, if the captain had delayed the order to shut down the No. 2 engine at the first indications of low oil pressure and quantity, he almost certainly would not have been able to restart that engine, and the aircraft would not have made it to land.

Damage to the Aircraft

Before getting to the cause of this near disaster, we will look at the damage to the aircraft. The only damage sustained was to the three engines.

The No. 1 and No. 3 engines sustained substantial damage. As just noted above, fire erupted in the No. 1 engine after landing, and smoke emerged from the No. 3 engine, indicating the start of a fire there. The No. 2 engine sustained minor damage, and was operable after servicing.

The Lockheed L-1011 was equipped with three Rolls Royce RB-211-22B turbofan engines. Each engine has six main rotating assemblies (plus an external gearbox) organized into seven major engine modules. [See Figure 3-2.]

All three engines were disassembled and examined by the National Transportation Safety Board (NTSB)—including draining of their high speed external gearboxes and oil tanks to determine the amount of residual oil in the gearboxes and oil tanks. The observed engine damage was consistent with loss of lubrication from loss of oil volume and/or loss of oil pressure.

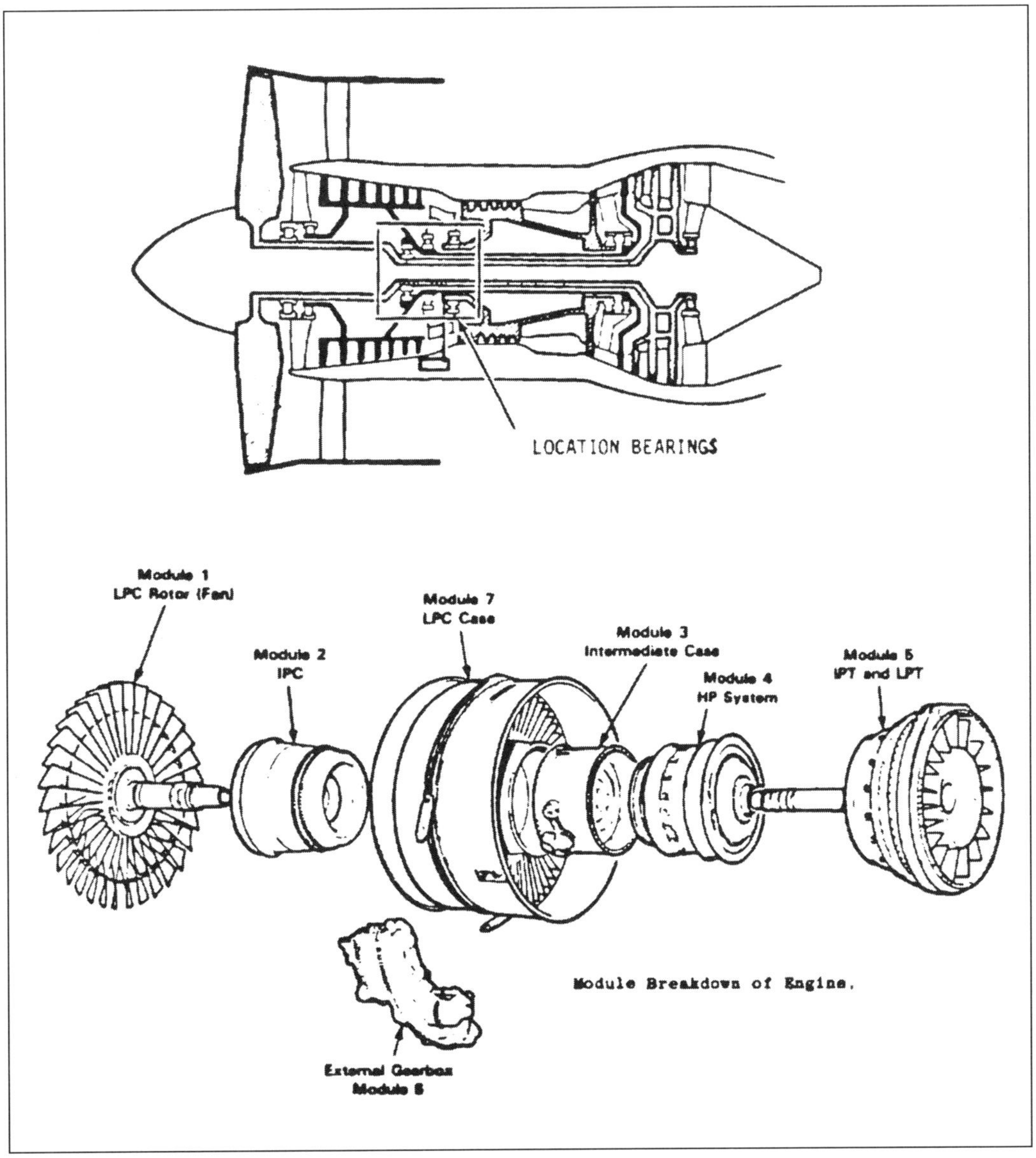

Figure 3-2: Engine Rotating Assemblies (top) and Module Breakdown (bottom). [from NTSB Report]

The No. 1 engine had a total of 5 quarts of oil remaining—4 quarts in the external gearbox and 1 quart in the oil tank. The No. 2 engine had 3 quarts remaining—2 quarts in the external gearbox and 1 quart in the oil tank. The No. 3 engine had 5 quarts remaining—3 quarts in the external gearbox and 2 quarts in the oil tank. Remember, for a "Full" indication, each engine should have had 21 quarts of oil.

No. 3 Engine Damage

Several components of the No. 3 engine were coated with a tacky oil film, resembling the appearance of freshly applied varnish. The intermediate pressure location bearing assembly and components were dry, severely overheated, and heavily discolored. There were heavy buildups of fused ball material on the rolling surfaces and two balls had large flats.

The high pressure location bearing had seized and could not be rotated. All of the balls in this bearing exhibited varying degrees of sliding and rotating flats.

The internal gearbox driven gear shaft had moved radially outward, making contact with the oil trough (baffle). One side of the baffle was broken.

The high pressure compressor driving bevel gear of module 4 was locked solid due to seizure of the high pressure location bearing.

The wear and deformation that resulted from the loss of lubrication caused misalignment of the rotating assemblies. The engine shut down because of a loss of gear mesh which disengaged the drive to the fuel pump.

No. 1 Engine Damage

Damage to the No. 1 engine was also significant, although not quite as extensive as that to the No. 3 engine. None of the components were covered with tacky oil as was found in the No. 3 engine. However, some components in the intermediate module were covered with a film of degraded oil.

The intermediate pressure location bearing assembly was dry externally and noisy when it was rotated. It was, however, easier to rotate than the No. 3 engine bearing.

The internal components were dry and discolored from the generated heat.

The high pressure location bearing assembly had seized and could not be rotated. The inner and outer races were dry and severely discolored from overheating. The bearing balls were similarly discolored and most exhibited wear flats and deformation.

The internal gearbox driven gear shaft had moved radially outward making contact with the oil baffle as in the No. 3 engine. This movement resulted in engine shutdown caused by loss of the fuel pump drive. The remaining components of the internal gearbox assembly were in a condition similar to that found in the No. 3 engine, except that the oil baffle was intact and undamaged, except for a light rotational rub.

No. 2 Engine Damage

Pre-repair testing of the No. 2 engine showed that all of the engine parameters (including vibration levels) were within prescribed limits.

There were some wear indications in the high pressure location bearing, but the remaining location bearings and components were in a normal condition.

The driven and driving gears of the internal gearbox assembly engaged normally and rotated freely. Disassembly showed some surface heat discoloration, but the bearing's components retained their original shape and dimensions.

Figure 3-3 shows representative examples of the kind of damage found in the engines.

Cause of the Occurrence

The immediate cause of this near miss event was—quite obviously—loss of lubricating oil from the engines. But the big question was what could possibly cause the loss of oil from all three engines? Particularly since each engine has its own independent lubricating oil system.

The answer was to be found in the routine scheduled maintenance performed on the aircraft during the night of May 4, 1983—mere hours before Flight 855 departed Miami International Airport on its scheduled flight to Nassau.

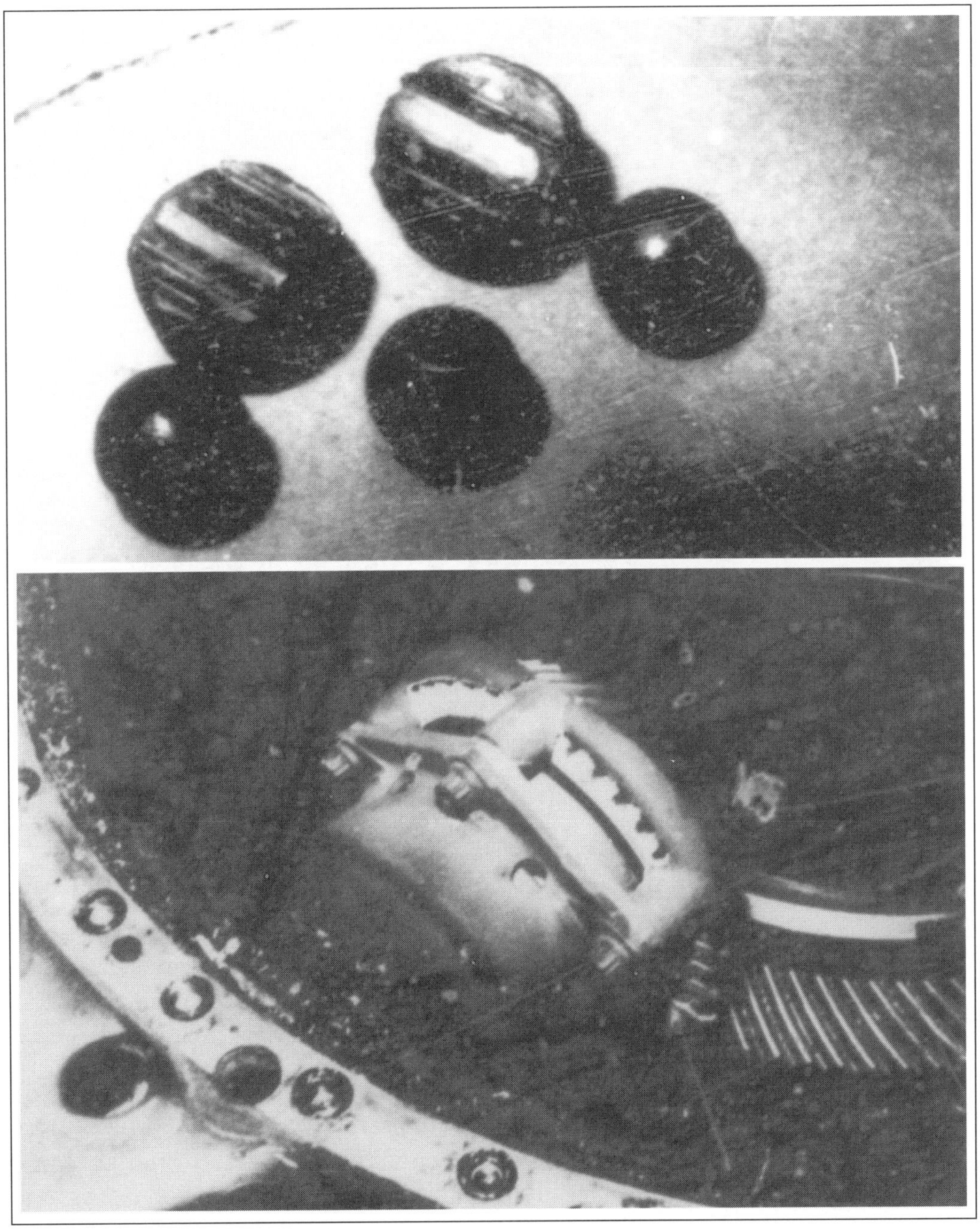

Figure 3-3: Engine Damage. Bearing balls showing sliding flats on two (top) and Internal Gearbox (No. 3 Engine) with displaced gearshaft (bottom). [From NTSB Microfiche]

Master Chip Detector Each of the engines has a master magnetic chip detector [Figure 3-4] positioned on the left side of the low pressure compressor case [see also Figure 3-2, module 7]. A magnetic probe which is part of the detector, protrudes into the oil line by opening a check valve and attracts small metal particles which may be present in the line. Metal particles in the oil indicate that internal components of the engine are wearing and that an engine may fail if the defect is not corrected.

O-Ring Seals Each master chip detector has two O-ring seals on the magnetic probe of the assembly as shown in Figure 3-4. These O-ring seals prevent oil from leaking from the pressurized oil system once the master chip detector is inserted in the engine, and displaces the check valve.

If a master chip detector is inserted into the oil system without O-ring seals, an oil leak will begin as soon as the engine is started. And that's exactly what happened with Flight 855. Mechanics failed to install any O-ring seals on all three master chip detectors. Note particularly that even one O-ring seal on a chip detector is sufficient to prevent oil leakage. Both seals must fail—or be left off—before oil will leak. Although no leak-rate tests were performed, about 15 quarts of oil leaked out of engines No. 1 and No. 3 in the accident aircraft in about 33 to 37 minutes.

Now, if a master chip detector falls out or is inadvertently not installed, the check valve in the oil line closes to prevent a serious oil leak. A small amount of oil will leak through the check valve, but, without a chip detector installed, it will take about five hours of operation before sufficient oil is lost to cause a low oil pressure warning.

Failed Maintenance

The maintenance performed on the Flight 855 aircraft was inspection and replacement of the master chip detectors for all three of the Roll Royce RB-211 engines. Eastern Airlines performed this maintenance routinely on all L-1011 aircraft after every 22 flight-hours—when an aircraft remained overnight at any of their maintenance stations having the capability to perform this check (Miami, Atlanta, New York, Boston, and Newark). If the aircraft was not scheduled into one of these maintenance stations, the inspection/replacement interval could go as high as (but not exceed) 40 flight-hours.

To perform this inspection, after a master chip detector is removed from an engine by the mechanic, it is examined by a technical field representative who evaluates the amount of metal particles on the detector. The results are coded and recorded for each engine, the technician cleans the detector and (in Miami) returns it to a cabinet in the foreman's office. If a substantial amount of wear particles are found, the detector is sent to a laboratory for analysis before being returned to the supply system.

Eastern Airlines required that only qualified mechanics be permitted to perform the master chip detector check. Eastern's Engineering Department estimated that more than 100,000 master chip detectors had been changed in the two years before this accident, and that each line maintenance mechanic involved in this occurrence had performed over 100 detector changes.

On May 4, 1983, the Flight 855 aircraft arrived at Miami where it was parked overnight at the Eastern Airlines terminal. The night-shift general foreman was responsible for completion of the scheduled work card tasks on the aircraft—which included the master chip detector inspection and replacement. Specific work assignments were made by a foreman.

Routine Work Card 7204 [Figure 3-5] was given to two mechanics to complete the master chip detector check. The procedure for this check is outlined in detail on the work card.

Performing the Maintenance

Let's review the work card and maintenance performed—one step at a time—to see if we can determine what caused the problems [Figure 3-5]. As you can see, it's a fairly simple, straightforward, six-step procedure.

Steps 1 and 2 involved simply removing the master magnetic chip detectors from all three engines and placing them in labeled containers for examination by the field technical representative. Those steps proceeded without incident. Figure 3-6 shows the access panels for the three engines.

Step 3 was to "reinstall/replace" the chip detectors in all 3 engines, with specific instruction to "use new seal" (O-rings) and "insure seal is properly installed and housing is in good condition". This, obviously, is where the maintenance failed. Although this work step is initialed as complete on the work card, no seals were installed on the detectors and (obviously) they did not insure that the seals were "properly installed". (We'll come back to this step later for a more detailed discussion.)

Figure 3-4: Master Chip Detector with O-Ring seals removed (top) and installed (bottom). [TECHSTAR]

Step 4, to check alignment holes to ensure correct installation of the probes was (as we shall discuss later) probably not performed—although the probes (albeit without seal rings) were properly installed.

Step 5 was to "motor engine and check chip detector for leaks". This step was performed, but was inadequate in its execution, as we shall discuss shortly.

Step 6, to close the access hatches for the three engines was correctly performed—although you will note that they failed to record the engine serial numbers in the spaces provided on the work card. (Note also that only one mechanic

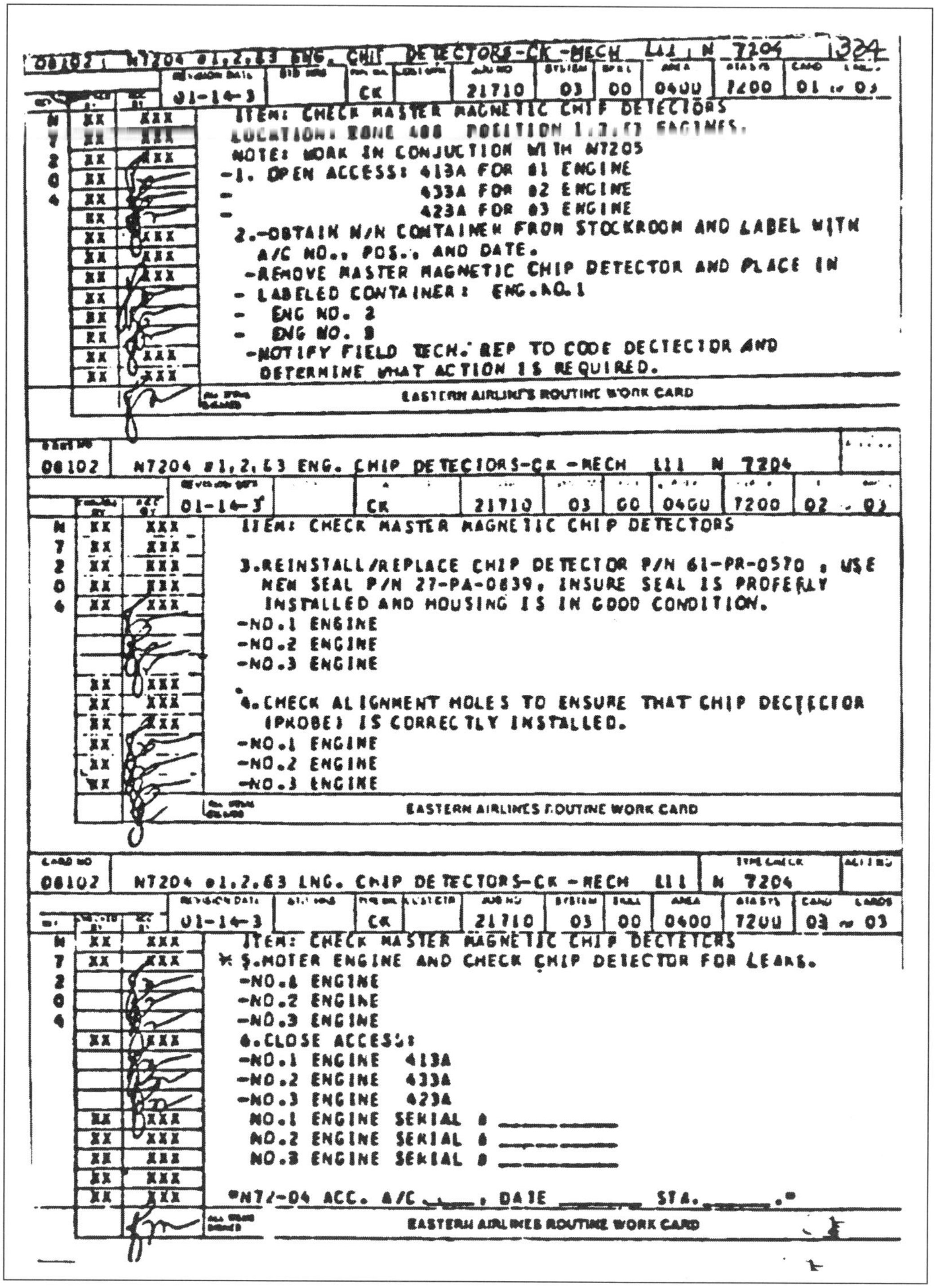
08102 N7204 #1,2,&3 ENG. CHIP DETECTORS-CK-MECH L11 N 7204 1324
01-14-3 CK 21710 03 00 0400 7200 01 of 03
ITEM: CHECK MASTER MAGNETIC CHIP DETECTORS
LOCATION: ZONE 400 POSITION 1,2,(3) ENGINES.
NOTE: WORK IN CONJUCTION WITH N7205
-1. OPEN ACCESS: 413A FOR #1 ENGINE
- 433A FOR #2 ENGINE
- 423A FOR #3 ENGINE
2.-OBTAIN N/N CONTAINER FROM STOCKROOM AND LABEL WITH A/C NO., POS., AND DATE.
-REMOVE MASTER MAGNETIC CHIP DETECTOR AND PLACE IN
- LABELED CONTAINER: ENG.NO.1
- ENG NO. 2
- ENG NO. 3
-NOTIFY FIELD TECH. REP TO CODE DETECTOR AND DETERMINE WHAT ACTION IS REQUIRED.
EASTERN AIRLINES ROUTINE WORK CARD

08102 N7204 #1,2,&3 ENG. CHIP DETECTORS-CK -MECH L11 N 7204
01-14-3 CK 21710 03 00 0400 7200 02 of 03
ITEM: CHECK MASTER MAGNETIC CHIP DETECTORS
3.REINSTALL/REPLACE CHIP DETECTOR P/N 61-PR-0570, USE NEW SEAL P/N 27-PA-0839, INSURE SEAL IS PROPERLY INSTALLED AND HOUSING IS IN GOOD CONDITION.
-NO.1 ENGINE
-NO.2 ENGINE
-NO.3 ENGINE
4.CHECK ALIGNMENT HOLES TO ENSURE THAT CHIP DETECTOR (PROBE) IS CORRECTLY INSTALLED.
-NO.1 ENGINE
-NO.2 ENGINE
-NO.3 ENGINE
EASTERN AIRLINES ROUTINE WORK CARD

08102 N7204 #1,2,&3 ENG. CHIP DETECTORS-CK -MECH L11 N 7204
01-14-3 CK 21710 03 00 0400 7200 03 of 03
ITEM: CHECK MASTER MAGNETIC CHIP DETECTORS
* 5.MOTOR ENGINE AND CHECK CHIP DETECTOR FOR LEAKS.
-NO.1 ENGINE
-NO.2 ENGINE
-NO.3 ENGINE
6.CLOSE ACCESS:
-NO.1 ENGINE 413A
-NO.2 ENGINE 433A
-NO.3 ENGINE 423A
NO.1 ENGINE SERIAL # ______
NO.2 ENGINE SERIAL # ______
NO.3 ENGINE SERIAL # ______
N72-04 ACC. A/C ____, DATE ______ STA. ______.
EASTERN AIRLINES ROUTINE WORK CARD

Figure 3-5: Work Card 7204. [from NTSB Report]

initialed all steps on the card as completed even though two mechanics performed the maintenance—one mechanic replaced the detector for two engines and a second mechanic replaced the detector on the third—No. 2—engine.)

Discussion of Failed Maintenance Steps

So, now we get to the "meat" of the failure. How could the mechanics have failed to properly perform such a "simple" task? (Note that alcohol, drugs, and fatigue were ruled out as contributors to this occurrence.)

Let's start with Step 3—the first failure—involving installation and inspection of the O-ring seals on the replacement probes. Looking again at the photo of the detector [Figure 3-4], it seems almost impossible not to observe (or even detect by feel) the omission of seal rings on the part. And the work card instructions specifically state that the mechanic is to personally install those seal rings.

The investigation and interviews with maintenance personnel and the involved mechanics revealed the following:

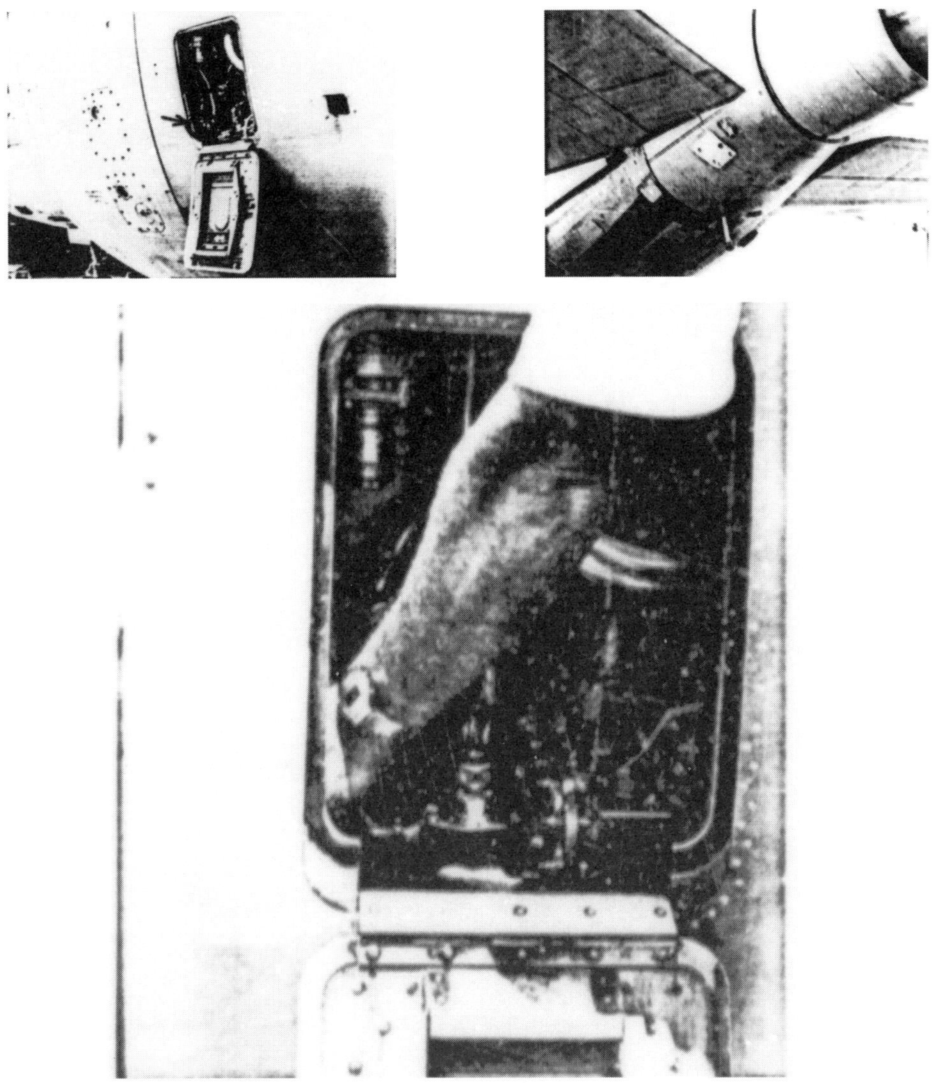

Figure 3-6: Chip Detector Access for No. 1, No. 2, and No. 3 (with mechanic reaching to detector location) Engines. [from NTSB Microfiche]

1. Mechanics performing this maintenance usually obtained the replacement detectors from a cabinet in the foreman's office, with new O-rings already installed.
2. On this occasion, there were no "replacement-ready" detectors in the foreman's office, and the mechanic obtained them from the stockroom. (Detectors in the stockroom have inspection tags, but do not have seal rings installed; that task, as indicated on the work card, is supposed to be performed by the mechanic himself at the time of installation.)
3. Poor illumination at the engine work site was cited by the workers as a factor making it more difficult to observe the condition of the probes, and installation of the detectors in the engines was done by "feel" rather than by direct observation.

Again, the work card clearly states that the mechanic will place new O-ring seals on the detectors before installation in the engines—and requires their signature (initials) verifying completion of that step. The mechanics themselves stated that they understood that requirement, but also stated that they thought the seal rings were being replaced by either the day shift or the technical field representative; they believed as a result that they did not have to replace the seals. Since they "always" got the detectors from the foreman's office with O-rings already installed, that assumption was continuously reinforced each night they performed work card 7204 that this is the way they come from maintenance supply. It became, therefore, normal practice for the mechanics to sign it off as complete even though they did not personally perform the task.

Supervisory Failure

Supervisory shortcomings contributed to this routine deviation from work card requirements. The general foreman stated that he knew that the mechanics were supposed to replace the seals, yet also admitted that he did not know of any mechanics who actually replaced the O-rings, and did not even know how or who installed them. So, the general foreman was aware that the published procedure was not being followed scrupulously, yet took no steps to correct the situation or to discover why the procedural deviation was occurring.

Lack of Light

The "lack of illumination" or "limited illumination in the maintenance area" (cited as "standard for the night shift") may have made proper inspection more difficult, but seems like more of an attempt at excuse than a legitimate "reason" for the failure. Should not good work practices dictate that the worker himself ensure that there's sufficient light to perform the work?

This claimed "lack of illumination" and installation of the detectors by "feel" also causes us to question whether Step 4—visually ensuring alignment—was actually or adequately performed.

Professional Care

The NTSB concluded:

> *The Safety Board believes that although the above factors might provide some rationale for the mechanics' actions, it does not excuse the mechanics' failures to adhere to the proper maintenance procedures. The mechanics were required to place o-ring seals on the detectors by Work Card 7204 and, as skilled and licensed mechanics, they should have inspected the detectors that they placed in the engine. Additionally, the work card was signed, signifying that new O-ring seals had been placed on the master chip detector when in fact they had not. As a result, the Safety Board concludes that the master chip detectors were installed without O-ring seals because the mechanics failed to follow the required workcard procedures, and because they failed to perform their duties with the professional care expected of an A & P mechanic.*

Take particular note of that last statement. They failed to perform their duties with the professional care expected.

Post Work Tests

On to Step 5. "Motor Engine and check chip detector for leaks." This step is essentially post-work testing—a final operational check to ensure that the detectors were properly installed and would not leak.

The mechanics did perform this step. They did motor each engine for 10 seconds, and subsequently checked each for oil leaks (finding none). Unfortunately, 10 seconds was not sufficient time to build up adequate oil pressure or to ensure that oil traveled to the chip detector location. (Subsequent tests indicated that a minimum 20 seconds of engine motoring would show any leaks; the procedure was thus modified to specify 30 seconds as the minimum motoring time for this check.)

This failure was not entirely the mechanics' fault. The procedure did not provide specific guidance on how long an engine should be motored, and they were not specifically trained on performance of this check. They only guessed that 10 seconds was sufficient.

The NTSB stated:

> *The lack of procedural guidance for the motoring check on work card 7204 was a deficiency in Eastern Air Lines maintenance supervision. The fact that the mechanics and the general foreman had no clear idea concerning the time necessary to accomplish the motoring procedure point to the fact that the maintenance personnel were given insufficient guidance and training and subsequently did not know how to perform the check properly.*

Excerpts from the NTSB conclusions and findings are:

- *Work card 7204 outlined proper procedures required to remove and reinstall a master chip detector, and to inspect the master chip detector for oil leaks.*
- *A master chip detector cannot be installed without O-ring seals if work card 7204 is followed.*
- *The mechanics had the responsibility to install O-ring seals and should have detected the absence of O-ring seals on the master chip detectors they installed.*
- *The mechanics experienced a deviation in their work pattern when replacement master chip detectors were not available from their normal source. The deviation did not prompt them to inspect the master chip detectors for the presence of the O-ring seals.*

Conclusions

So, what are we left with? Other than insufficient guidance and training for the post work leak test, the cause lies solely with the mechanics' failure to properly perform a quite simple, routine inspection/maintenance task—one that they both had performed hundreds of times before. And, their failure to install six small O-rings almost cost the lives of 172 people.

Procedures for the maintenance performance were certainly clear, straightforward, and sufficient to preclude error.

And, don't forget, there were two mechanics performing this work on this aircraft—both of them failed. One mechanic replaced the detectors on the No. 1 and No. 3 engines and the second replaced the one in the No. 2 engine.

We are left with two (admittedly skilled) mechanics who both failed to perform a simple routine task. How could this possibly happen?

The NTSB states:

> *The investigation of the accident revealed several factors which may have influenced the actions of the mechanics. The first factor was the frequency with which master chip detectors are changed and the ease in which the installation is completed. "Routineness" of any task can lead to unquestioning, inattentive action and, in this accident, both mechanics demonstrated a lack of attention to the procedures and an unquestioning assumption that the detectors came with O-ring seals installed.*

So, as always, the bottom line—the last line of defense against disaster—lies with reliance on each individual worker. Reliance on each worker to perform assigned tasks accurately, professionally, and paying attention to the details.

Learning from Flight 855

In **The Industrial Operator's Handbook** we stated that "there is no substitute for the alert, well-trained operator controlling equipment parameters within specified operating bands in accordance with approved procedures". That tenet applies equally for maintenance where the control of equipment parameters involves restoring and maintaining equipment in accordance with its original operating/configuration specifications—and strictly adhering to approved procedures in the performance of that work.

Proper vigilance, proper professionalism, and proper attention to the task at hand are no less important in performing maintenance than for operating the equipment. The mechanic's job is just as important as the pilot's job.

Well-Trained is Not Enough

In **The Industrial Operator's Handbook** [Chapter 6, The Alert, Well-Trained Operator], we delineated the general requirements and processes for selecting and training staff, both for maintenance and operations. Yet, no amount of training can compensate for a lack of alertness, whether it is the result of fatigue, illness, or complacency. In the case of Flight 855, the National Transportation Safety Board found nothing deficient about the training of the mechanics involved in the event. They simply did not execute their task with the vigilance expected of aircraft mechanics.

Have a Procedure—And Follow It!

Having good procedures available is part of the strategy for operating success. But just having the procedure isn't good enough. *Using* the procedure is also necessary. In this event, the procedure was clear and straightforward. However, the mechanics who performed the task simply initialed the step for installing O-rings without actually verifying that the step was complete.

Heed Others' Errors

This was not the first event in which O-rings had been left off of master chip detectors. Just in Eastern's maintenance history, there had been a number of similar occurrences, though none that resulted in the nearness to an accident as aboard Flight 855. The only value in identifying and reporting near misses is to educate others who are likely to find themselves in similar circumstances. At least for these two mechanics, their leaders had failed to sensitize them to the recurrent error trap of leaving O-rings off chip detectors.

Post Work Testing

The purpose of post work testing is to verify that maintenance has been successfully completed and that equipment has been restored to design specifications.

Eastern's post work test procedure for master chip detector replacement was correct except that it didn't specify how much time the engines should be motored in order to build sufficient pressure to see if the chip detectors leaked. And, apparently, that information was not a part of the mechanics' knowledge base.

Case Study References

NTSB National Transportation Safety Board. Aircraft Accident Report—Eastern Air Lines, Inc., Lockheed L-1011, N334EA, Miami International Airport, Miami, Florida, May 5, 1983. Report Number NTSB/AAR-84/04, March 9, 1984.

National Transportation Safety Board. Investigation File Documents, NTSB File Number MIA83FA136, [Microfiche 23663-A].

Chapter 4

Integrity of the Envelope

Maintenance is the process of restoring equipment to design specifications—bands of tolerance which prescribe the composition and performance of parts, components, and systems. The purpose of *corrective* maintenance is to return degraded equipment to design requirements when parts or components fail. The purpose of *preventive* maintenance is to detect and forestall degradation in equipment so that design specifications are not breached.

Since safety envelopes are developed based upon prescribed equipment performance limits, maintenance plays a vital role in protecting the integrity of envelopes. If safety-related materials and components are allowed to degrade beyond acceptable limits, protection cannot be guaranteed. Consequently, maintenance team members have as much to do with safety envelopes as do operators, builders, and designers.

In this chapter we will examine the importance of maintaining equipment within specified bands as we review a classic example of failed maintenance—the allision of the bulk freighter *Bright Field* with Poydras Wharf near New Orleans on December 14, 1996.

Bright Field Freighter Allision, December 1996

On December 14, 1996, a fully loaded bulk ocean freighter, navigating down the Mississippi River near New Orleans, experienced an unexpected engine shutdown caused by low pressure in the main engine lubricating oil system. Prompt restart of the engine was impeded by transfer of engine control from the wheelhouse to the engine room and subsequent poor communication between the two stations. Without power, the pilot was unable to maintain steering control. The ship's downriver momentum carried it at a ground speed of approximately 10 mph toward a wharf crowded with cruise ships, a river excursion boat, and a casino paddle-wheeler, all of which were laden with passengers. Although engine power was soon restored, the pilot was unable to re-establish directional control since the ship was in a "rudder stall" condition. When an allision (the marine term for a ship hitting a stationary object) with the wharf seemed imminent, the engine was reversed, but forward momentum could not be arrested. The *Bright Field* struck the Poydras Street Wharf, narrowly missing one cruise liner and the casino paddle-wheeler, causing extensive damage to the wharf and the Riverwalk shopping complex. Sixty-seven people were injured, some seriously, while trying to escape the wharf area and during the rapid evacuation of the vessels alongside the pier. Damages to the *Bright Field* totaled nearly $2 million. Poydras Wharf and the Riverwalk shopping complex suffered an additional $15 million in damages.

Root Cause Following the accident, a marine board of investigation convened by the United States Coast Guard determined that the *Bright Field* disaster was directly due to poor maintenance of the main engine and its associated lubricating oil system—problems that had plagued the freighter for months and which were well-known not only to shipboard officers, but also to leaders of the operating company:

> *The root cause of the casualty was ineffective management and oversight of the vessel's machinery condition and operational and maintenance practices. Long standing and recurrent problems with the main engine were not adequately diagnosed or remedied by shipboard officers. Shoreside managers failed to pay sufficient attention to the machinery performance and maintenance reports to assure that problems were corrected and that the vessel would perform safely.*

Contributing Factors Contributing to the accident were errors in communication, procedure non-compliance, ignorance of warning signs, a general failure in bridge resource management (command and control) aboard *Bright Field*, the pilot's lack of familiarity with the freighter's handling characteristics, and, from a broader perspective, inadequate management oversight and support on the part of the China Ocean Shipping Company (COSCO).

Troubled Voyage

The *Bright Field* is a 735-foot dry bulk ocean freighter powered by a single 5-cylinder 9,800 horsepower direct drive diesel engine. Licensed in Liberia, it was owned by Clearsky Shipping Company and operated by COSCO out of Hong Kong. The vessel's engine room was certified by Det Norske Veritas (DNV) as a "Periodic Unattended Machinery Space" using automated engine-operating controls, allowing the engine room to be periodically unattended during normal operations.

Figure 4-1: Bright Field Being Towed Away From Poydras Wharf. [From NTSB Report]

Det Norske Veritas (DNV) is an independent foundation in Oslo, Norway, providing marine vessel reliability, classification, and certification services. Ships navigating in United States waters must meet the rules and regulations set forth by the DNV for power plant reliability.

Bright Field's westward voyage began in Banjarmasin, Indonesia, on September 12, 1996, where she loaded coal destined for the United States. [See Figure 4-2.] Her schedule included a short layover in Singapore before proceeding through the Straits of Malacca, around the southern tip of India, into the Gulf of Aden and the Red Sea, through the Suez Canal and the Mediterranean, then across the Atlantic and into the Gulf of Mexico to discharge her cargo of coal in Davant, Louisiana, just short of New Orleans on the Mississippi River.

History of Engine Problems

During the voyage to Singapore, *Bright Field*'s crew was forced to stop the vessel at sea three times—once on the 13th, again on the 17th, and for four more days beginning the 18th—to inspect the power plant and perform emergency maintenance. The engineering staff found cracked cylinder liners, damaged pistons, broken and damaged piston rings, and fouled fuel injectors. Additionally, the main engine lubricating oil purifiers and strainers were not performing effectively, lending to a problem with high particulate concentrations in the lube oil. Repairs included fuel injector replacement, piston ring replacement, and cylinder liner weld repairs. Because of an inadequate spare parts inventory, the chief engineer had to utilize scavenged parts, reuse parts that had previously been removed from the engine, and make temporary weld repairs. Limited power was restored and the *Bright Field* continued to Singapore at reduced speed. *Bright Field*'s master informed COSCO that, upon arrival in Singapore, in-port repair of the main engine would be necessary.

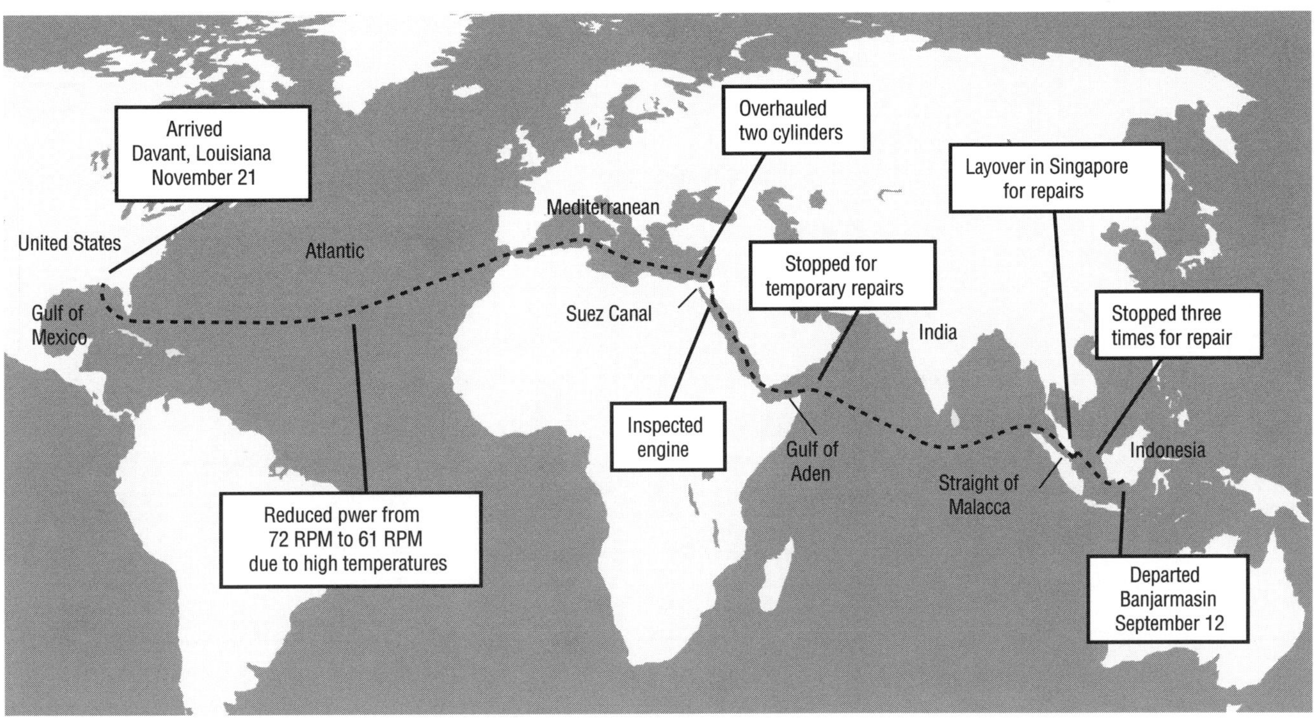

Figure 4-2: Bright Field's Route. [TECHSTAR]

Bright Field's plague of engine problems was not new. She had a long history of slow downs and stoppages due to various engine component and control failures. The list of problems included cracked cylinder heads and liners, damaged pistons, carbon-fouled fuel injectors (at a frequency much higher than normal), broken piston rings, burned exhaust valves, cracks in cylinder water jackets, and fuel pump problems—indicators of excessive engine temperatures, ineffective cooling, improper fuel-air concentrations, incomplete combustion, and inadequate engine lubrication.

A Month Behind Schedule

The scheduled 8-hour stopover in Singapore extended to a multi-day repair outage, delaying the *Bright Field*'s departure for the Suez Canal until early in the month of October. En route to the Suez, the freighter again had to be stopped for repairs when the engine developed high exhaust temperatures. Temporary repairs were performed and the *Bright Field* resumed her voyage. COSCO was again notified of the engine's condition and repairs made.

Bright Field arrived at the Suez Canal on October 21, nearly a month behind schedule. While awaiting passage, the freighter's engineering staff pulled the diesel's cylinder heads, inspecting the heads, cylinders, pistons, and rings. Finding damage, they informed COSCO that repairs would be required in Port Said after passage through the canal. After transiting the canal, the *Bright Field* lay in port on October 23 and 24 as the crew overhauled two of the engine's cylinders and made minor repairs to one of the other cylinders. The crew determined that poor quality piston rings were contributing to the repeated engine failures and notified COSCO.

Repairs at Sea

But *Bright Field*'s engine problems had not been resolved. On November 5, while crossing the Atlantic, the crew had to reduce power from 72 RPM to 61 RPM in response to high turbocharger exhaust temperature. Later on the same day, a lubricating oil alarm sounded requiring the engineers to shut down the engine. Suspecting fuel-contaminated lube oil, the crew changed the oil in the sump with approximately 2,220 gallons of new oil from the bulk storage tank, expending their store of engine lubricating oil. Following repairs for the high exhaust temperature problem, *Bright Field* was able to resume her voyage on November 7.

Over the next several days the crew encountered continuing power plant problems, some requiring engine shutdowns to perform maintenance. On November 14, *Bright Field*'s officers, believing their freighter could not complete her voyage without extensive repair, requested COSCO to arrange for engine and fuel system maintenance upon arrival in New Orleans. She limped into the Gulf of Mexico under reduced power, and began navigating up the Mississippi River to Davant, Louisiana.

Arrival In U.S.

Replacement of Chief Engineer

The *Bright Field* arrived at a bulk coal handling facility in Davant, Louisiana, on November 21, where she unloaded her cargo of coal over the next four days. Concurrently, the chief engineer was removed by COSCO for being "unresponsive to the owners' orders" according to the National Transportation Safety Board accident report. The new chief engineer came aboard and initiated engine repairs. *Bright Field*'s crew was assisted by a representative from New Sulzer Diesel (manufacturer of the *Bright Field*'s main engine) who inspected the engine and investigated the reasons for ongoing failures. He noted the following deficiencies:

- Cracked cylinder liners;
- Missing, worn, dirty, carbon-contaminated components, including piston rings;
- Clogged cylinder inlet ports and piston cooling passages;
- Clogged, dirty, damaged components to charge air receiver; water leaks observed;
- Fuel oil pump drain clogged and leaking fuel oil into main engine lubricating oil sump (recommended testing oil as soon as possible);
- Missing main engine turbocharger filters and dirty internals;
- Control air system valves clogged with condensation;
- Worn, loose, misaligned fuel oil pumps and linkages.

Repairs at Davant

While anchored near Davant, the *Bright Field*'s crew began cleaning the cargo holds in preparation for a load of corn. They failed their first hold inspection on the 25th. In the meantime, engine maintenance was performed including fuel injector replacement, cylinder and piston repair work, fuel pump overhauls, and fuel injector timing. Also, the bulk oil storage tank was replenished with approximately 2,000 gallons of engine lube oil.

Deteriorating Crew Morale

Bright Field finally passed cargo hold inspection on December 5 and moved upriver. On December 6, she arrived at the Bonnet Carre Anchorage, located approximately 10 miles downriver from the Reserve, Louisiana, loading center. [See Figure 4-3.]

Then on December 7, in preparation for leaving Bonnet Carre to transit upriver to Reserve, main engine testing identified a high turbocharger inlet temperature necessitating repair of both the turbocharger and air cooler. The master informed

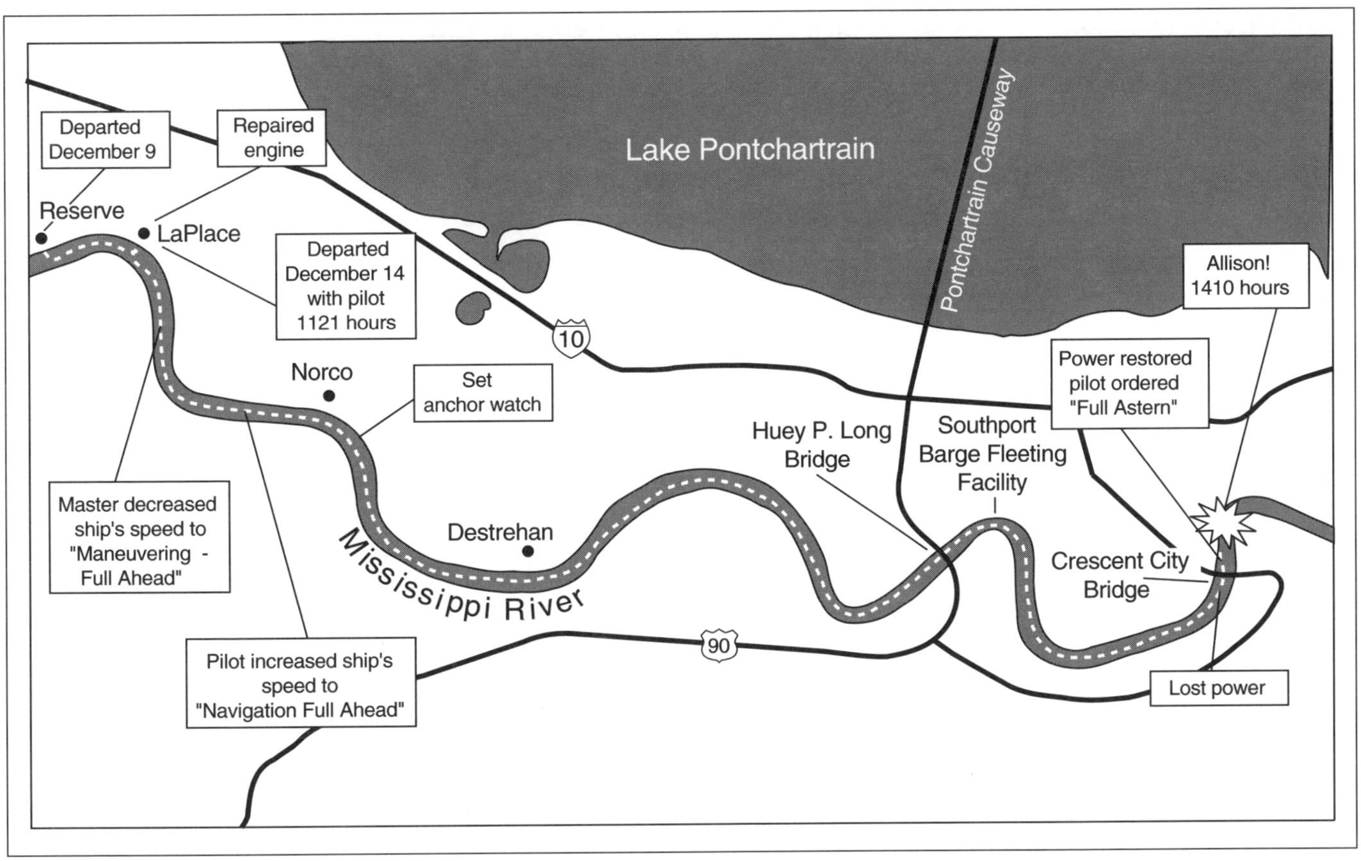

Figure 4-3: Bright Field's Route from Bonnet Carre Anchorage. [TECHSTAR]

COSCO of the *Bright Field*'s continuing engine problems and also told the owners that most of the crew had applied for leave because of deteriorating moral. An English translation of the master's message to COSCO reads:

> *Because the main engine of the Bright Field still has problem(s), and the process of washing holds did not go smoothly, while COSCO Hong Kong has explicitly expressed that the crew will not have any bonuses, the crew members are joining together to ask for leave. The leadership of the vessel and each department, under the circumstances, cannot do anything but their best to maintain order, and to make sure that everybody finishes his own work. We can only expect the Company to handle this accordingly.*

The ship remained at anchor until December 9 when it was moved to the Louis Dreyfus Grain Elevator at Reserve. The chief engineer noted that both oil pumps were "noisy" and reported that he would conduct further checks on the pumps at a later date. Engine inspections revealed worn turbocharger rotor rings and problems with the #5 cylinder piston rings.

On December 11, the *Bright Field* finished loading 56,380 tons of corn and moved to the LaPlace Anchorage for additional engine maintenance. While anchored at LaPlace, the turbocharger and air cooler were removed and taken to a shore maintenance facility for cleaning and repair as scheduled by COSCO. The turbocharger and cooler were returned and reinstalled on December 13. The engine was started and appeared to operate without difficulty.

Returning to Sea

Preparation for Downriver Navigation

On December 14, *Bright Field* was scheduled to navigate back down the Mississippi River to begin her voyage to Cristobal, Panama. Her estimated arrival date in Cristobal was December 19. The Mississippi was at high river stage, almost four feet over normal level with a 4 ½ mph current. Regulated Navigation Areas (RNAs)—a Coast Guard-controlled navigation aid for river safety employing red and green passage lights—were in effect. In anticipation of the *Bright Field* returning downriver, the lights were switched to green for downriver navigation, and deep draft ships going upriver were held below New Orleans. Shallow draft vessels were allowed to continue operating upriver provided they gave way to deep draft vessels. A pilot was contracted through the New Orleans Baton Rouge Steamship Pilots Association (NOBRA) to guide the *Bright Field* downriver to New Orleans. A second pilot would take the ship from the trade off point to the gulf.

Pre-Departure Checks

In preparation for a 10:00 A.M. departure, *Bright Field*'s crew began performing pre-departure checks of the freighter's propulsion unit and controls, steering and navigation systems, and auxiliaries. Under the direction of the new chief engineer, the *Bright Field*'s engineering staff tested the main engine lubricating oil pumps as well as salt water and fresh water cooling pumps, placing the main pumps in service and the backup pumps in standby mode, according to the chief engineer's testimony to the Coast Guard's Marine Board of Investigation. (In later testimony, however, the chief engineer admitted placing the second main engine lubricating oil pump in manual start mode rather than standby mode.)

Important ship's support systems such as main engine lubricating oil are typically serviced by two pumps—one running and one in standby. The pump controllers usually have three positions: RUN, STANDBY, or OFF. The standby feature on the main lube oil pump controller is designed so that the standby pump will start if the running pump fails or ceases to provide output of a minimum value. If the motor controller is placed in the OFF (manual start) position instead of the STANDBY position, the pump will not start when the control circuitry detects that the running pump is failing. In manual mode, an operator must take a manual action to start the pump.

With the help of the ship's electrician, alarm lights were tested; but alarm signals from the engine room to the wheelhouse were not tested. At 9:43 A.M., the *Bright Field* was readied for its downriver transit.

Pilot Aboard

The pilot arrived on board *Bright Field* at 10:40 A.M., having inspected the ship's draft and list before boarding. He met with the master, finding his English broken, but understandable. The second mate and the helmsman were able to repeat back commands in English, and the pilot stated that he was satisfied with the bridge crew's ability to communicate.

The pilot asked the master if all navigational equipment and the ship's engine were functioning properly, and he was assured that they were ready. The pre-departure discussion, however, was not as thorough as it was supposed to be. The Coast Guard Marine Board of Investigation Report states that:

> *The discussion did not include a review of methods or procedures for Pilot-crew communication or bridge management during an emergency, nor did it involve a review of the vessel particulars, draft, maneuvering characteristics, or vessel peculiarities as required by 33 Code of Federal Regulations 164.11. Moreover, the Pilot looked at the vessel particulars and maneuvering information posted on the bridge, but he testified that he did not thoroughly review the available information on the vessel's propulsion plant, nor the stated maneuvering or stopping performances, prior to getting underway.*

Even though this was his first time aboard *Bright Field*, the Pilot did not adequately familiarize himself the ship's equipment and handling characteristics as required by law.

Indication of Engine Problems

At 10:55 A.M. the ship was ready to get underway. The pilot's first command after the anchors were hoisted was, "Dead Slow Ahead." The third mate attempted to execute the pilot's order using the wheelhouse engine controls, but the engine failed to start. The third mate contacted the engine room by ship's telephone and, speaking in Chinese, transferred control below to the chief engineer. After the chief engineer started the engine, he transferred control back to the wheelhouse. The master later testified that this was normal practice in non-emergency situations.

The pilot then ordered the engine to "Stop", allowing the ship to drift out of anchorage. While the ship was clearing the anchorage, the pilot again ordered, "Dead Slow Ahead," and again the engine failed to start from the wheelhouse. Once more, control was transferred back to the engine room. The engine room maintained engine control due to a problem with a fuel control valve, but after about 20 minutes control was again transferred back to the wheelhouse. Since all communications from the wheelhouse to the engine room were in Chinese, the pilot was unaware that engine control was being transferred back and forth. The pilot was not informed of the inability to start the engine from the wheelhouse, and no engine alarms sounded on the bridge to alert the pilot to the change in control.

Underway

To familiarize himself with the ship's rudder response, the pilot ordered "Maneuvering Full Ahead" (56 RPM). The ship responded as expected, so the pilot then ordered the freighter to sea speed (72 RPM). He found the ship handled better at higher speed. He then ordered the third mate to reduce speed to "Maneuvering Full Ahead" as they approached Norco, Louisiana. The pilot maintained engine revolutions at 56 RPM until reaching Destrehan. Each engine power change was accomplished at the wheelhouse without transfer to the engine room.

After reaching Destrehan, the pilot ordered the engine speed increased to 72 RPM, propelling the *Bright Field*, in conjunction with river current, at approximately 16 knots ground speed. As required by the New Orleans Port Authority, the pilot ordered the anchor watch set. The master directed the ship's carpenter to stand by the anchors with a two-way radio, ensuring that they were held only by the friction brake for rapid deployment if necessary.

Bright Field encountered river traffic in the vicinity of Nine Mile Point, first meeting a cargo ship and then a tug and tow. As per river navigation procedures, VHF communication was established and both pilots agreed on how to safely pass each other. The cargo ship passed without incident, but the tug and tow were off course, necessitating evasive action by *Bright Field*.

Engine Room Alarms

From the time the *Bright Field* left anchorage to the passing of the two vessels, numerous alarms sounded in the engine room—none of which were known to the pilot. They included:

- Engine lubrication second filter—high differential pressure (alarmed 4 times)
- Cylinder cooling water—high temperature across all 5 cylinders (alarmed 2 times)
- Scavenging air—high temperature
- Main engine cooling sea water—low pressure (alarmed 2 times)
- Number 1 lubricating oil purifier—abnormal condition

In each case, the alarms were acknowledged and reset by the engineering staff in the engine room (or automatically reset after the alarm condition cleared). The Coast Guard recorded, "No alarms, indicating a problem with the main engine, sounded on the Bridge on 14 December 1996." Moreover, neither the master nor the pilot was notified of the alarm conditions, the causes of the alarms, or the actions taken to clear the alarms.

Freighter Out of Control

As the *Bright Field* passed under the Crescent City Connection Bridges, another alarm sounded. The main engine lubricating oil low pressure alarm activated and tripped the engine off line. At this point, *Bright Field* was a little over a thousand yards from its allision point with Poydras Wharf, a distance that would be covered in 3 minutes and 25 seconds.

Loss of Power

The bridge crew felt a reduction in the ship's normal operating vibrations. The second mate noted that the engine tachometer reading had fallen from 72 to 30 RPM. The pilot asked the master, "Do we have a problem?" Without replying, the master ordered the second mate to call the engine room to determine what happened and to bring the engine back to maneuvering speed. The pilot did not understand the master's communications since the master and second mate conversed in Chinese. No reports were rendered to the pilot in English, nor did the pilot demand such reports. Yet, in testimony, both the master and second mate stated that they had responded to the pilot.

Observing a loss of the main engine, the second mate contacted the chief engineer stationed below in the engine room. They agreed to transfer engine control to the engine room to restore power. Testimony indicates that the second mate communicated with the chief engineer in a normal tone of voice, perhaps without imparting a sense of urgency and danger. Again, though engine control had once more been transferred to the engine room, the pilot was not informed.

Loss of Steering

Bright Field's bow was now swinging toward port. The pilot ordered the rudder hard to starboard in an effort to regain directional control. The helmsman complied with the order, but the *Bright Field* did not respond to the rudder change. Accident investigation later determined that the ship was probably in "rudder stall"—a condition in which the rudder is ineffective in providing directional control due to flow disturbance along the rudder paddle. [See Figure 4-4.]The Coast Guard reports that:

> *Thomas C. Gillmer, in his book, Modern Ship Design (Naval Institute Press, Annapolis, Maryland, 1986), defines stall as the sudden discontinuity of lift on the downstream surface of the rudder caused by an increasing angle of attack to the critical angle where separation occurs and when normal flow pressures can no longer exist…If propeller RPM rapidly decreases or the propeller is put in reverse, the volume of flow from astern of the propeller decreases, thereby decreasing the flow onto the rudder. Simultaneously, a temporary excess in volume on the forward side of the propeller pushes the streamlines out away from the rudder which increases the angle of attack. If the rudder is put over to the stops, it also pushes out the streamlines on that side, increasing the angle of attack, contributing to rudder stall and loss of steerageway.*

Sounding the Alarm

With the *Bright Field* out of control and heading toward a wharf crowded with people and ships, the pilot's first reaction was to notify the Governor Nicholls Light operator, hoping that the operator could warn the ships and businesses at Poydras Wharf: "70 to Governor Nick: I just lost my engine. Alert everyone in the harbor to watch out!" The light operator notified local ships, contacted the Marine Safety Office New Orleans, and requested emergency tug assistance.

As the *Bright Field* closed with the wharf, the pilot again called the Governor Nicholls Light operator saying, "Governor Nick, I'm going to hit that damn cruise ship. Call him or do something. Governor Nick, call him, or something, I'm goin' right for him!" On a collision course with the *Queen of New Orleans*, a gambling ship moored at Poydras Wharf, the pilot radioed the light operator once more, "Warn the gambling boat I'm going right for him!"

The pilot began sounding *Bright Field*'s whistle in a series of five short consecutive blasts (indicating a ship in distress) to warn vessels and people of the impending allision.

Engine Power Restored

Below in the engine room, the engineering staff was trying to determine the cause of the main engine trip. The audible alarm was acknowledged and the alarm readout was reviewed to determine the failure. The readout revealed three activated alarms:

- *M/E LO IN PRESSURE A[T] 2.49 KG/CM2. (The main engine lubricating oil pressure had dropped below the alarm pressure of 2.5 kg/cm².)*
- *M/E MAIN LO PRESS TRIP. (The main engine had automatically shut down due to low lubricating oil pressure at the main bearing.)*
- *M/E AUTO CONTROL FAIL. (The main engine control system was disabled due to an abnormal condition and the main engine would not respond to speed or direction commands from the bridge or the engine room control consoles.)*

The automatic control system first detected low oil pressure to the main engine bearings. Next, to protect the engine, the auto controller shut down the main engine. Finally the controller locked out control from both the wheelhouse and en-

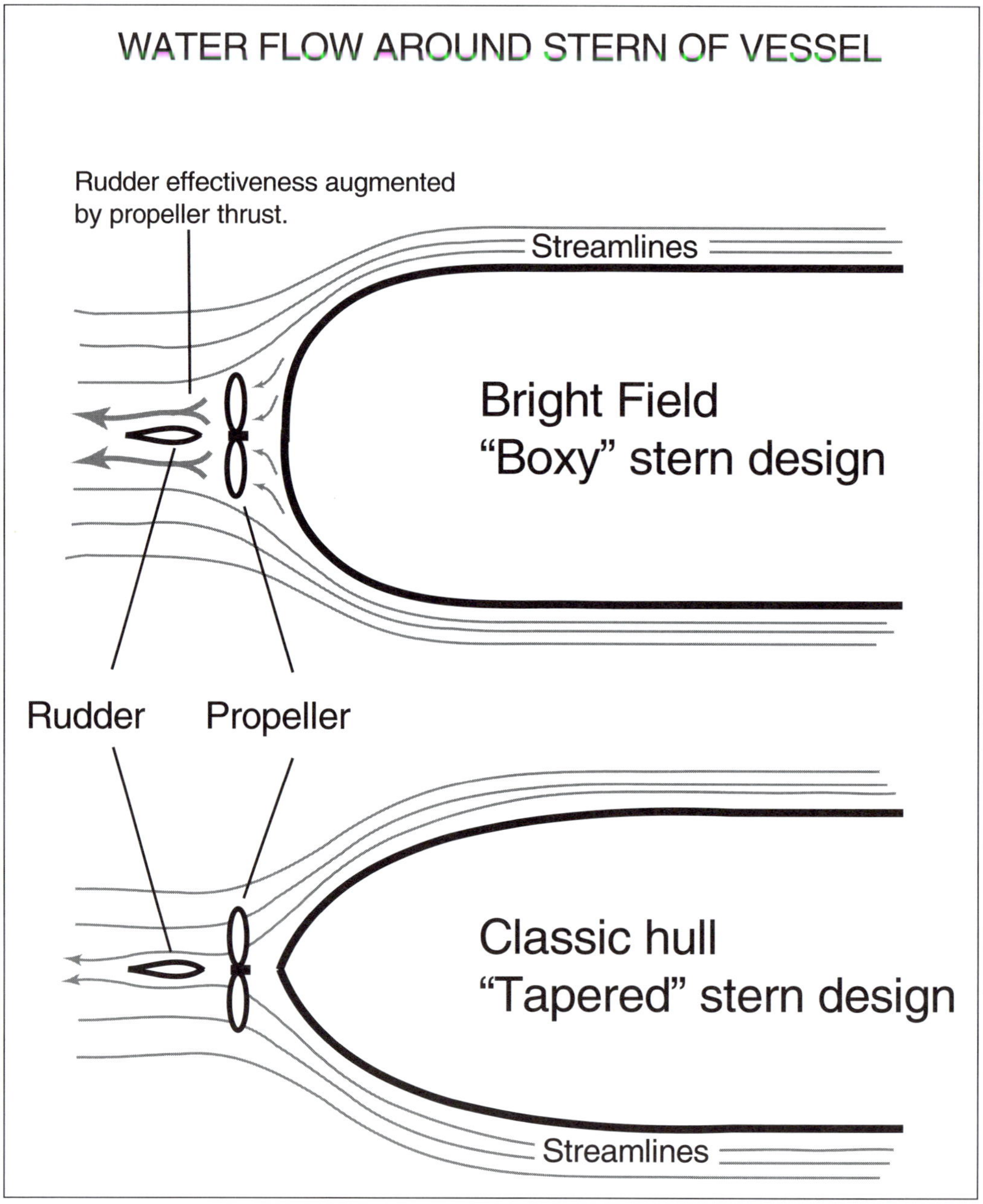

Figure 4-4: Bright Field Hull Showing Water Flow. Because of the "boxy" hull design, when the ship is not under power, water does not flow around the stern to the rudder (as it does with a "tapered" design) and steering control is lost. Also, with this configuration, when the rudder is at an extreme angle, the rudder will "stall" as described in the text. [TECHSTAR]

gine room requiring a manual reset of the trip. This feature prevents the controller from automatically restarting the engine without crew intervention.

Determining that low lubricating oil pressure was the cause of the main engine trip, the chief engineer checked the lube oil pumps for operation. The # 1 pump was running, but the pressure was below the low trip setting. The chief engineer stated in testimony that, as he was checking # 2 pump, it started on its own, promptly restoring main engine lubricating oil pressure above its trip set point. After correcting the oil pressure problem the chief engineer restarted the engine. (In later testimony, the chief engineer recanted his testimony, admitting that the standby lubricating oil pump was, in truth, in the OFF position at the time of the problem. It had to be started manually at the time of the low lube oil pressure emergency.)

The *Bright Field* was equipped with an emergency maneuvering override feature, the purpose of which was to bypass normal engine speed-limiting functions for emergency situations. Certainly, the *Bright Field*'s loss of propulsion at a critical time warranted use of the override feature. Yet, it was not activated. The National Transportation Safety Board reported that:

> *The scavenging air pressure limiter functioned to reduce main engine exhaust emissions. Overriding the limiter injected more fuel into the engine, thereby increasing engine rpm at a slightly faster rate. The limiter was automatically overridden whenever wheelhouse controls were moved to "crash (emergency) ahead" or "crash astern." Also, a crash maneuvering pushbutton on the engine control room console enabled engineering personnel to override the scavenging air pressure limiter if necessary. According to the second engineer, the crash maneuvering pushbutton was used only on the orders of a bridge officer, and such an order was not given on the day of the accident.*

Asked why the emergency maneuvering feature wasn't used, the master stated that it was because "...the engineering department [didn't] realize the sense of urgency or [that] there [was] any sense of urgency on the bridge."

Allision Imminent!

Once the pilot determined that an allision with the wharf was unavoidable, he began to focus on controlling the impact. He remained uninformed of actions taken by the crew to correct the loss of power; however, when engineering restored power he noted the increase in engine revolutions on the tachometer.

Both the pilot and the master concluded that it was too late to control the ship's direction using forward power. To slow the ship's momentum, the pilot directed the Master to prepare to drop anchors and the second mate to bring the engine to "Full Astern".

Below in the engine room, as the chief engineer attempted to restore ship's power back to 72 RPM, he encountered additional alarm and control conditions which limited the acceleration to about 60 RPM. And, just as forward speed was being established, the bridge ordered the engine to "Full Astern". The engineers were still unaware *Bright Field* was about to allide with Poydras Wharf.

Engine control remained in the engine room with the chief engineer rather than in the wheelhouse. Though the chief engineer was still unaware of the emergency situation, he apparently was beginning to sense the urgency of the situation based upon the series of commands and engine orders he was receiving. When the engine order telegraph was rung to "Full Astern", he employed engine starting air as a brake to the diesel's forward rotation. Then, once the engine was stopped, he reversed it and brought speed up as quickly as possible. Stopping and reversing the engine took approximately 30 to 40 seconds after the order was received; but, again, the crash maneuvering override feature was not employed.

As the pilot sounded the ship's whistle to warn those on and near the wharf, he also ordered the master to drop the port anchor. The master tried to contact the anchor watch using his two-way radio, but the noise from the ship's whistle prevented the anchor watch from hearing the order. The master then attempted to gain the anchor watch's attention by waving to him from the port bridge wing. By the time communication was established, the master determined that dropping the anchor at that point would increase the likelihood of striking the *Nieuw Amsterdam*, one of the cruise ships anchored at the wharf. So, without informing the pilot, he chose to delay execution of the order. Then, just prior to contacting the wharf, the master ordered the anchor released. Seeing that the ship was about to allide with the wharf, the anchor watch partially released the brake and left his station. The port anchor never made it to the waterline.

Striking the Wharf

At 2:11 P.M., *Bright Field*'s bow struck Poydras Street Wharf at a 45-degree angle. The bow penetrated the wharf nearly 60 feet, and then moved sideways down the wharf. The *Bright Field* fortuitously missed the passenger liner, *Nieuw Amsterdam*, and the casino ship, *Queen of New Orleans*, coming to rest between them. Several structures were severely damaged, including a parking garage, the Riverwalk shopping mall, and part of the Hilton Riverside Hotel.

Three vessels and a crowded wharf lay in the freighter's path. [See Figure 4-5.] The pilot of the *Bright Field* notified the Governor Nicholls Light operator, asking that all the ships and shore facilities be warned. Affected vessels along the wharf were the *Nieuw Amsterdam* (a cruise liner), the *Queen of New Orleans* (a casino ship), and *Creole Queen* (an excursion boat).

Poydras Street Wharf

On the wharf, most people crowding the stores and restaurants were unaware of the emergency situation until moments before impact. Public safety officers attempted to warn wharf patrons and direct them away from the impact area. Clerks and patrons were ignorant of the meaning of the ship's warning whistle and did not expeditiously evacuate. Shortly after the allision, several New Orleans emergency units arrived to provide crowd control, secure the damaged area, treat the injured, and search for unaccounted and trapped persons. Some units were delayed in reaching the injured next to the river due to damage caused by the allision.

Nieuw Amsterdam

Nieuw Amsterdam's third mate, stationed on the ship's bridge, heard the *Bright Field*'s warning whistle and immediately notified the master. The master ordered embarkation of passengers to be stopped at the gangway and for all watertight doors to be secured. Pier side passengers were directed into the terminal building. The master then ordered all passengers on the lido deck evacuated.

Queen of New Orleans

The first mate on the *Queen of New Orleans* had been monitoring VHF channel 67 and overheard the *Bright Field* call the Governor Nicholls Light operator. The first mate notified the master who ordered an evacuation. Emergency evacuation announcements were made and the general alarm was sounded. The master ordered the chief engineer to start the engines and prepare to get underway. Passengers were issued life jackets and casino employees were instructed to evacuate the ship after the passengers.

An obstacle to quick evacuation of the *Queen of New Orleans* was the use of a single gangway creating a bottleneck. Passengers jammed into the area next to the gangway, slowing the evacuation. The bow wave preceding the *Bright Field* caused the gaming boat to "bounce" up and down, requiring the crew to secure the gangway until it was safe for passengers to cross again. Approximately 80 percent of the passengers were safely evacuated.

The master ordered the crew to search for passengers left onboard and search for anyone who may have fallen over the side. The crew then stood by to help emergency personnel on the wharf.

Creole Queen

The master of the *Creole Queen* was monitoring VHF Channel 67 while he was preparing to get underway. After hearing of the emergency he suspended departure and initiated an evacuation. He announced the evacuation over the public address system and sounded the general alarm.

The incoming bow wave from the *Bright Field* caused the *Creole Queen* to pull away from the pier, dropping the gangway and three people into the water. Life rings and jackets from the boat were thrown to the people in the water below. Two of the overboard passengers climbed up a ladder lowered down to them and the third passenger was rescued using a Stokes stretcher. Once the bow wave subsided, the gangway was pulled back into place and the rest of the passengers were evacuated.

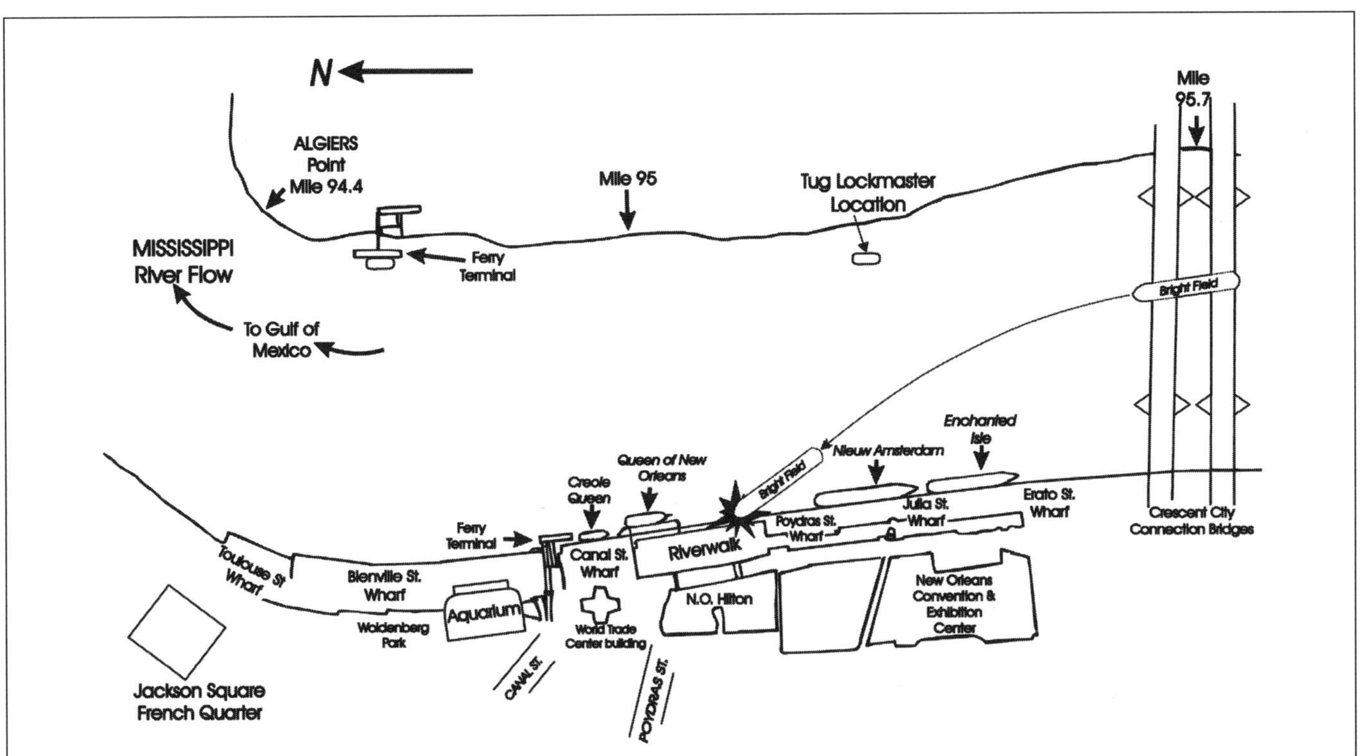

Figure 4-5: Path of the Bright Field to Poydras Street Wharf. [From NTSB Report]

The Cost

The *Bright Field*'s allision injured 67 people, 4 seriously. Damage to the *Bright Field* was estimated in excess of $2 million. More than 350 feet of wharf frontage was damaged or destroyed. Fifteen businesses, including shops, restaurants, and hotels were either temporarily or permanently closed. According to the NTSB, wharf and business damages exceeded $20 million. Civil litigation costs were termed "incalculable" by a local newspaper. The *Bright Field* remained at the Poydras Wharf from December 14, 1996, to January 6, 1997. Tugboats then moved it to the Bollard Marine Repair facility near Violet, Louisiana. In late February, the Coast Guard released the ship to leave U.S. waters.

The Coast Guard found the *Bright Field*'s crew negligent in vessel operation. Citations included failure to notify the Coast Guard of hazardous operations, failure to inform the pilot of hazardous ship conditions, and failure to properly test communications and engine alarm systems before departure. Subsequently, COSCO was fined $800,000, the maximum allowable for the infractions. Further, the Coast Guard notified the Liberian Bureau of Maritime Affairs (LBMA), the licensing agency for the *Bright Field*'s officers. The Coast Guard Marine Board made this sobering recommendation:

> *The Republic of Liberia should conduct a suspension and revocation investigation of the BRIGHT FIELD's licensed engineers for negligence and misconduct while under the authority of their licenses and documents in that they apparently: failed to maintain an appropriate level of lubricating oil in the main engine sump, failed to properly set the #2 main engine lubricating oil pump for automatic operation and provided false sworn testimony to the Marine Board.*

Causes and Contributing Factors

The Domino Chain

The immediate cause of the *Bright Field* event, obviously, was a loss of engine power due to a main lube oil low pressure trip. The Coast Guard Marine Board wrote:

> *The proximate cause of this casualty was the loss of main engine lubricating oil pressure to the main bearings which resulted in a main engine trip, causing the loss of propulsion and steerageway.*

But other conditions, actions, and omissions—other dominos in the chain—made the situation worse. Coast Guard investigators continued:

> *The sequence of RPM loss, hard right rudder and full astern propeller all reduced water flow over the rudder, induced a hydrodynamic stall and prevented the Pilot and Master from regaining positive control of the BRIGHT FIELD.*

In other words, loss of power led to loss of steering control—an emergency made worse by an ineffectively coordinated (and poorly executed) emergency response.

Candidates for Cause

In **The Industrial Operator's Handbook, 2nd Edition** [Chapter 2, Common Components of Accidents], we identified twenty of the most common accident contributors. In the *Bright Field* event, prominent contributors include:

- Deficient equipment maintenance—the main engine and main lube oil system (as well as associated alarm and control systems) were not maintained in accordance with technical specifications.

- Procedure non-compliance—failure to maintain equipment in accordance with manufacturers' specifications necessitated violation of New Sulzer procedures, Terasaki procedures, Det Norske Veritas procedures, and COSCO's own procedures.
- Ignorance of warning signs—the maintenance history of *Bright Field*'s main engine and supporting systems provided ample warning that the ship was in an unreliable state, an especially dangerous condition for operating in restricted waters.
- Poor leadership and supervision—the bridge crew of *Bright Field* (pilot and master, in particular) did not employ the precepts of proper *bridge resource management* designed to limit errors of command and control due to poor communication and incomplete understanding of responsibilities of the members of the bridge crew.
- Incomplete communication—*Bright Field*'s pilot was not kept informed of the status of the main engine, especially of whether control remained at the wheelhouse or in the engine room; nor was the chief engineer and his staff made aware of the seriousness of *Bright Field*'s situation at the time the emergency occurred.
- Inadequate understanding of equipment operating characteristics—the pilot was not fully aware of *Bright Field*'s handling characteristics, especially her tendency to rudder stall because of a less streamlined stern design.
- Insufficient management oversight—COSCO failed to provide the maintenance support and oversight necessary to keep *Bright Field* in reasonable running condition (another illustration of poor leadership).

Note the mixture of both operating and maintenance factors. Maintenance mistakes triggered the event, but operating responses then exacerbated the situation. We see again that operations and maintenance are closely related functions. The actions of one ultimately affect the performance of the other. Note also that the influence of leaders resides at the fulcrum in the balance between success and failure. Where leaders weigh in is often the crucial difference.

Underlying Causes

Each of the factors just identified has one or more underlying causes. The questions that cause analysts must answer are, "Which of the contributing components had the greatest influence in causing the accident?" and, "What were the underlying causes of those factors?"

In the *Bright Field* event, deficient maintenance tops the list of contributors. The main engine and lubricating oil system had reached such an unreliable state that consequences (beyond the cost of repetitive repairs) were probably inevitable. Coast Guard investigators determined that the maintenance deficiencies were not simply the result of errors on the part of *Bright Field*'s staff. Rather, there was a deeper, underlying cause:

> *The root cause of the casualty was ineffective management and oversight of the vessel's machinery condition and operational and maintenance practices. Long standing and recurrent problems with the main engine were not adequately diagnosed or remedied by shipboard officers. Shoreside managers failed to pay sufficient attention to the machinery performance and maintenance reports to assure that problems were corrected and that the vessel would perform safely.*

In other words, *Bright Field*'s and COSCO's longstanding habits of deficient maintenance practices finally caught up with them—at a most inopportune time.

Deficient Engine Maintenance

Since this book is about maintenance and its relationship to the safety envelope, let's focus for a moment on how *Bright Field*'s equipment degraded to the state of repair in which she was found on the day of the incident.

Both the Coast Guard and the National Transportation Safety Board determined that *Bright Field*'s engine tripped off line because of low lubricating oil pressure. But even a cursory review of *Bright Field*'s main engine and lube oil system history indicates that this event was not an anomaly. Rather, it was an event "under construction"—an error-likely condition created by poor maintenance practices.

Recurring Engine Problems

The U.S. Coast Guard inspected *Bright Field* and its maintenance records in June of 1996. At that time all maintenance items had been cleared. Perhaps the inspections were not thorough, or perhaps the *Bright Field* began experiencing a series of maintenance problems after the inspection. Regardless, the National Transportation Safety Board found that:

> *The Bright Field showed evidence of recurring engineering problems that affected vessel main engine reliability, and had all engineering systems been kept in good repair and regularly tested, the vessel may not have unexpectedly lost power during its voyage down the Mississippi River.*

Here is a monthly summary of known engine repairs performed subsequent to the annual Coast Guard inspection:

June (post inspection)
- Replaced #2 cylinder liner with used liner in better condition
- Repaired crack in #3 cylinder head by welding
- Replaced 5 of the 10 fuel injectors believed to be carbon fouled
- Replaced fuel pumps on cylinders #4 and #5
- Replaced exhaust valve in #1 cylinder

July
- Numerous slow downs and stops for engine maintenance recorded in ship's log – repairs unknown
- Removed, cleaned, and tested all fuel injectors
- Replaced 4 fuel injectors
- Replaced piston rings on cylinders #4 and #5 (#4 rings were previously used)

August
- Replaced both fuel injectors on #4 cylinder
- Repaired water leak on #5 cylinder

September
- Replaced #1 cylinder liner with previously used liner – failed to repair by welding
- All 10 fuel injectors replaced
- Overhauled air starting valves on cylinders #1, #3, #4, and #5
- Replaced fuel injectors on cylinders #3 and #5
- Replaced #4 cylinder liner with previously used liner (new liner not in stock)
- Reconditioned #4 piston
- Removed #5 cylinder liner, piston, and connecting rod – resumed voyage at reduced speed
- Repaired #4 cylinder liner, piston, and air cooler
- Replaced #1 piston rings
- Replaced #5 cylinder sleeve and piston – upon arrival at port

October
- Replaced fuel injectors on cylinders #1 and #3

Dismantled and cleaned 2 engine speed control valves
Overhauled cylinders #2 and #3, renewed rings; found scavenging box dirty with heavy deposits of carbon

November
Cleaned scavenging box
Replaced fuel injectors on cylinders #4 and #5
Repaired piston crown on #2 cylinder
Replaced high pressure fuel pumps on all 5 cylinders
Adjusted fuel injector timing
Overhauled #5 cylinder – installed new cylinder liner
Replaced fuel injectors on cylinders #1, #2, and #3
Renewed 3 air diaphragms in main engine emergency stop valve

December
Turbocharger and main air cooler cleaned and repaired

This litany shows that engine component failures aboard *Bright Field* were significantly higher than normal. Typically, fuel injectors are replaced every 8,000 hours of use. Aboard *Bright Field* engineers were replacing injectors at an average of 920 hours. Piston rings are designed to last 15,000 hours before replacement. Rings in *Bright Field*'s main engine were only lasting 780 hours.

The majority of these problems stemmed from improper fuel combustion (causing excessive cylinder temperatures) and poor engine lubrication. Two common contributors to improper fuel combustion are incorrect fuel injector timing and restricted air intakes, both of which were determined to be problems aboard *Bright Field*. And an important side effect of improper fuel combustion is the buildup of carbon deposits on engine components and carbon contamination of lubricating oil.

Ineffective Preventive Maintenance

Upon arriving in Davant, Louisiana, on November 21, the engine manufacturer's representative boarded and began diagnosing *Bright Field*'s engine problems. He found the injection system improperly timed, cylinder fuel pumps improperly installed, and air cooler and air exchange boxes dirty and restricted. Before leaving he corrected the fuel timing, re-installed the fuel pumps, and recommended cleaning the air coolers and scavenging boxes.

The ship's engineering staff had worked on each of these items before reaching Louisiana, but their efforts had been ineffective. The engineering staff may not have understood the relationship between the multiple main engine problems—poor fuel combustion, lube oil contamination, and accelerated component failure. Part of the problem was also related to the loss of a critical tool. Investigators learned that the crew had lost or misplaced a specialized tool used for fuel injector timing. Without the use of the correct tool, fuel timing adjustments were questionable.

The problems discovered by the New Sulzer representative are also indicative of a failure to perform effective preventive maintenance. The National Transportation Safety Board recorded that:

> *While the Bright Field's owners provided each vessel in their fleet with general requirements for periodic testing and maintenance of the main engine, including regular analysis of the main engine lubricating oil, Safety Board investigators determined that critical main engine components were run until failure occurred and that periodic preventive maintenance was not routinely accomplished.*

Without an effective preventive maintenance program in place, *Bright Field*'s engine problems had compounded. The *run it 'til it breaks* philosophy of maintenance had, once again, proven to be exorbitantly expensive.

Contaminated Lube Oil

One of the most egregious maintenance shortcomings of *Bright Field*'s engineering staff was the failure to maintain the engine lubricating oil system in accordance with procedure and manufacturers' specifications. The Coast Guard Marine Board wrote:

> *Continued operation of the main engine using lubricating oil which did not meet the engine manufacturer's specification for viscosity and water content, and containing excessive particulate contamination is an indication of improper engineering plant operation and inadequate maintenance.*

Investigators also found that:

> *Analysis of the used lubricating oil indicated a failure of the engineers to monitor and maintain the main engine lubricating oil.*

Samples of engine lube oil from the oil sump in November, 1996 (before the accident), and on January 5, 1997 (after the accident), were tested for viscosity and water content. Results concluded that both viscosity and water content exceeded the manufacturer's guidelines. The Coast Guard wrote:

> *The viscosity of two of the four samples were outside the allowable +/-10% difference from new oil while all four samples exceeded the 0.2% maximum allowable water content, with one sample being 7.6% water.*

Had the crew conducted the recommended oil particulate tests, they would have discovered high levels of carbon contamination also.

At least five problems underlay *Bright Field*'s main engine lube oil difficulties: 1) failure to maintain the main engine in a state that allowed proper and efficient combustion of fuel, 2) failure to adequately operate and maintain the lube oil purifier and lube oil filters and strainers, 3) failure to clean and inspect the oil sump prior to refilling with new oil, 4) failure to maintain sufficient oil level in the main engine sump, and 5) failure to keep important lube oil parameter sensors and alarms within manufacturer's tolerances. A combination of these problems contributed to the low lube oil pressure trip of *Bright Field*'s main engine at a critical juncture in their transit down the Mississippi River.

From Inefficient Combustion

As you have probably learned from years of automobile operation, inefficient combustion of fuel ultimately leads to particulate contamination of the lubricating oil in an engine through carbon buildup and excessive erosion of engine components. The *Bright Field* was experiencing an exaggerated form of that problem. The Coast Guard Marine Board wrote:

> *A contributing cause [of the main engine trip] was the failure of COSCO and the vessel's crew to properly address long standing problems (at least six months duration) with the main engine which led to the high level of particulate in the lubricating oil.*

Particulates from inefficient fuel combustion clogged filters and screens in the lube oil system, resulting in restricted oil flow.

From Improper Oil Changes

Part of *Bright Field*'s oil contamination problem resulted from failing to follow procedure and good maintenance practices for changing oil in the sump. After leaving the Suez Canal the crew suspected that the lube oil was contaminated with fuel, so engineers pumped the used engine oil from the engine sump and replaced it with 2,219 gallons of new oil from the storage tank. This action, though intended to resolve the contaminated oil issue, simply cross-contaminated the new oil.

Procedure required the main engine oil sump to be cleaned and inspected after the oil was drained. An industrial diesel engine representative (from New Sulzer) believed as much as 10 percent of the oil's contaminants remain in the sludge on the sump wall and bottom after draining used oil. Therefore, it is essential to clean the sump to prevent new oil from becoming contaminated. The representative also pointed out the need to inspect a clean and empty sump for screen and mechanical damage and other abnormalities since they cannot be seen when the sump is full. *Bright Field*'s engineering staff neither cleaned the sump nor inspected it before refilling.

From Clogged Filters and Strainers

Because of particulate, emulsified oil and water sludge, and sediment in the oil sump, lube oil filters and strainers aboard *Bright Field* frequently exceeded differential pressure limitations, a condition which can cause pump cavitation and frothing of oil—problems which affect lubrication. After the accident, the second filter in *Bright Field*'s lube oil system was examined and found to be clogged with particulate and sludge. Instead of keeping the filters clean (and differential pressure within specification), the crew simply reset the filter's high differential pressure alarm to a value 30 percent higher than manufacturer's specifications. After the accident (and before resuming operation), *Bright Field*'s lube oil system required thorough cleaning to restore filter and strainer function to specification.

From Low Oil Level

Bright Field's lube oil cleanliness problems were aggravated by the engineering staff's failure to keep the oil sump filled to the minimum level specified by the technical manual. The freighter's main engine oil sump is designed to hold 2,721 gallons of 30 weight high detergent lubricating oil. Factory specification requires a minimum quantity of 2,536 gallons. Maintenance of this minimum oil level in the sump is important since, as oil cycles through the lubricating system, it picks up air. The more time the oil has to circulate through the system, the more time entrained air has to escape. Entrained air reduces oil's ability to lubricate moving parts. Reducing the sump's oil level by 317 gallons reduced the oil's circulation time from 6 to 3 minutes, reducing the oil's efficiency. According to COSCO's technical advisor, it was permissible to use purified used oil as make-up if there was an inadequate supply of new oil; nevertheless, *Bright Field*'s main engine oil sump had run for long periods at a low level.

From Improperly Adjusted Set Points

To accommodate low lube oil sump level without subjecting themselves to an annoying low lube oil level alarm, *Bright Field*'s crew had adjusted the lube oil alarm set point about one-fourth below manufacturer specification. The Coast Guard Marine Board wrote:

> *A contributing cause was the low volume of lubricating oil in the main engine sump at the time of the casualty. The main engine sump low level alarm was operating but had been adjusted to allow operation below the normal alarm set point without sounding the alarm.*

The oil sump low level alarm is normally set at 2,536 gallons, but at the time of the accident the sump only contained 1,982 gallons. Investigation found the low level alarm set at only 1,849 gallons or 27 percent below manufacturer's minimum prescribed level. Periodic oil sump soundings (manual level measurements) recorded the level at only 1,982 gallons. As stated earlier, low oil level leads to an excess of entrained air. Accident investigators learned that air entrainment prevented the lube oil pump from maintaining pressure.

The sump oil level alarm was not the only instrument out of calibration, though. The Coast Guard Marine Board wrote:

> *A contributing cause was that an improperly calibrated lubricating oil pressure sensor prevented the Terasaki monitoring and alarm system from alerting the engineers in the engine room that lubricating oil pressure was dropping and a main engine trip was imminent.*

The *Bright Field* was outfitted with a Terasaki monitoring and alarm system—part of the equipment that qualified the ship to operate the engine room as a "Periodic Unattended Machinery Space" at certain times (in unrestricted waters). One of the engine's operating parameters monitored by this system was main engine oil pressure. But the oil pressure sensor for this system was found to be out of calibration in the low direction. The Coast Guard Marine Board wrote:

> *...the alarm printout from 14 December 1996 clearly showed the main engine tripped due to the loss of lubricating oil pressure before the low lubricating oil pressure alarm sounded. This is contrary to proper functioning of the automation system which was designed to sound the alarm before the main engine trip. The alarm is designed to sound when the pressure reaches a value of 2.5 kg/cm^2 and the trip occurs when the value drops to 2.3 kg/cm^2.*

The main engine trip function sensed from a different source than the alarm function. With the alarm sensor out of calibration in the low direction, the main engine tripped before the low pressure alarm sounded. As a result, *Bright Field*'s crew had no alarm warning of dropping main lube oil pressure before the main engine tripped. As a result, the main engine trip was probably a bigger surprise than it otherwise would have been.

Deficient Maintenance Support

Clearly, the corrective and preventive maintenance programs aboard *Bright Field* were mismanaged and ineffective. But excellent maintenance at the operating level is unlikely if the maintenance philosophy at the managing level is lax. As the operator for Clearsky Shipping Company, COSCO was contracted to operate and maintain the *Bright Field* (as well as several other vessels in Clearsky's fleet) in a professional, workman-like manner. Yet, both the Coast Guard's and National Transportation Safety Board's investigations indicate that maintenance support and advice at COSCO's level was deficient. The National Transportation Safety Board noted that:

> *The Bright Field's owners' oversight of testing and maintenance of the vessel's engineering systems was inadequate and led to unreliable performance of the engineering plant and contributed to the shutdown of the main propulsion engine on the day of the accident.*

As we learned earlier, the *Bright Field* had a long history of major engine problems and repairs. The recurrence of engine problems indicates that repairs were little more than temporary. At least from July of 1996, *Bright Field* was in need of professional maintenance assistance. COSCO leaders knew this. The National Transportation Safety Board wrote:

> *The Bright Field's owners, Clearsky Shipping Company, received regular engineering and maintenance reports on the Bright Field and thus had knowledge of the vessel's engineering problems. Serious engineering problems were reported to the owners as early as January 1996. In addition to the reports from the engineering crew, the owners had received periodic telexes from the master documenting the severity of engineering problems, delays in scheduling resulting from the engineering problems, and the inability of the engineering crew to make permanent repairs…These engineering deficiencies are indicative of long-term, recurring problems that cannot be tied to the competence or job performance of a single individual.*

In other words, the ship's problems were not simply the result of a poor chief engineer. Nor would replacement of the chief engineer solve the *Bright Field*'s extensive main engine problems.

COSCO's poor logistical support is also illustrated by *Bright Field*'s inadequate spares inventory—lack of new parts for engine repair—and, at least in the case of piston rings, substandard parts. The engineering staff was forced to reuse parts that had been previously discarded and to make weld repairs of cylinder sleeves instead of being able to replace the cylinder sleeves.

COSCO also did not ensure that *Bright Field*'s first chief engineer (from Indonesia to the U.S.) was thoroughly trained and competent. Investigators determined that he did not have the ability to correctly direct the complex repairs, nor did he recognize the causes of the multiple engine failures.

Factors Affecting Command and Control

Without question, poor maintenance underlay the loss of control of *Bright Field* on the day of the event. But, as we have already suggested, other factors—especially poor command and control on the bridge of the freighter—made an undesirable situation worse. The National Transportation Safety Board wrote:

> *Training in BRM [Bridge Resource Management] typically includes five generic categories of knowledge and skill development: (1) the development and performance of watch or pilot briefings; (2) maintenance of situational awareness; (3) identification of error chains (and error trapping); (4) implementation of effective bridge/vessel communication; and (5) integration (coordination) of bridge/vessel resources. In whole or in part, most of these elements were missing during the Bright Field's December 14, 1996, voyage.*

Let's examine a few of *Bright Field*'s missing BRM pieces.

Incomplete Familiarization by Pilot

Before departure, river pilots are required by federal code (33CFR164.11) to become acquainted with their ship's handling characteristics, communication conventions, and emergency procedures. Aboard *Bright Field*, the pilot's familiarization was limited. In particular, his knowledge of the freighter's engine and rudder response were acquired as the freighter left it's anchorage in LaPlace.

The design of the *Bright Field* was optimized for maximum cargo capability, fuel efficiency, and for passage through the Panama Canal. The *Bright Field*'s "boxy" design creates streamlines (water flow around the hull) that do not force water across the rudder. Rudder effectiveness is directly dependent on propeller RPM. The faster the engine is run, the greater the rudder response. More classic "tapered stern" designs draw water across the rudder at all speeds.

In this incident, when the engine stopped, the flow of water over the rudder was greatly reduced, diminishing the rudder's effectiveness. At that point the pilot ordered full starboard rudder. The rudder went into "rudder stall", pushing the remaining streamlines away from the rudder and creating water "turbulence" on the backside. Subsequently, the freighter lost steerage.

Deficient Departure Briefing

Closely related to the pilot's familiarization is a pre-departure briefing—a discussion prior to movement to establish ship's condition, command and control, communication conventions, transit plans, and emergency response actions. The departure briefing aboard *Bright Field* was superficial. Coast Guard investigators wrote:

> *As the Pilot was familiarizing himself with the bridge, he and the Master briefly discussed the vessel's equipment. The discussion did not include a review of methods or procedures for Pilot-crew communication or bridge management during an emergency; nor did it involve a review of the vessel particulars, draft, maneuvering characteristics, or vessel peculiarities as required by 33 Code of Federal Regulations 164.11. Moreover, the Pilot looked at the vessel particulars and maneuvering information posted on the bridge, but he testified that he did not thoroughly review the available information on the vessel's propulsion plant, nor the stated maneuvering or stopping performances, prior to getting underway.*

One of the elements missing from the briefing was a downriver transit plan. The Mississippi was at high river stage and the Coast Guard had Regulated Navigation Area restrictions in effect. This required the pilot and master to pay special attention to the red and green passage lights. The pilot and master did not discuss the procedures for passing river traffic, what speed the ship would transit down the river, or what allowances would be necessary to compensate for the high water.

Nor were emergency response actions discussed. The National Transportation Safety Board wrote:

> *A transit plan should include intended speeds, areas of high risk due to traffic concentration, shoreside structures, or river flow characteristics, and actions to be taken in the event of various ship power or control difficulties. According to their statements, each man felt confident of his own abilities and assumed that the other was qualified to perform any duties that might be required. The lack of discussions between the master and pilot regarding emergency maneuvering procedures did not cause this accident. Still, if the actions to be taken in an emergency had been established, cooperation and coordination in the emergency that developed would probably have been enhanced, tasks could have been clearly delineated, and orders (such as the order to drop anchor) could perhaps have been given and complied with more quickly.*

Communication conventions are an important part of both normal and emergency operations. When engine power was lost on the bridge and control was shifted to the engine room, the pilot was not advised of the change. Had the pre-departure briefing included instructions for notification of the pilot when engine control was changed, he would have been aware that his commands would be delayed while being relayed to the engine room. The third mate would also have understood the necessity for reporting his actions to the pilot. Instead, the pilot was ignorant of the ship's procedure for managing bridge control failures and whether the ship's telephone or engine telegraph would be used for relaying engine commands in an emergency.

Incomplete Communication

Successful command and control—the object of bridge resource management—depends in great part upon concise and meaningful communication. On *Bright Field*'s bridge, communication failed.

At the first engine command from the pilot to the third mate, the engine failed to start from the bridge. The third mate did not tell the pilot that he was unable to control the engine from the bridge and had to transfer control to the engine room. The same problem occurred at the pilot's next engine command and, again, the third mate did not inform the pilot of the problem. Since the transfer of engine control was conducted in Chinese, the pilot was unaware of the problem. Consequently, he took the *Bright Field* downriver without knowledge of the engine control deficiency, a dangerous lapse in ship handling.

The pilot's ignorance of equipment status continued when the *Bright Field* lost engine power, and the pilot asked the master, "Do we have a problem?" The master did not answer and the pilot did not inquire again. As a result, the pilot was making decisions based on his assumptions rather than knowledge of fact.

A third example of ineffective communication on *Bright Field*'s bridge occurred when, as engine control was transferred to the engine room following loss of power, the second mate did not inform the chief engineer of the nature of the ship's emergency. The engineers learned of the impending allision just moments before contact with the wharf.

Procedural Non-Compliance

Procedural non-compliance—both in maintenance and operations—contributed materially to *Bright Field*'s incident. One of *Bright Field*'s operating procedures required that, prior to getting underway, one main lube oil pump should be placed in RUN and the other in STANDBY. As we learned before, should the primary pump fail to establish minimum lube oil pressure, the other would automatically start. Yet, in violation of procedure, the #2 pump was placed in the OFF position and, therefore, did not automatically start—a precursor to loss of main engine power.

Another procedural violation—failure to maintain and test bridge alarms—resulted in the master and pilot being deprived of important engine status information during a critical juncture in the emergency. Investigation revealed that engine alarms on the bridge were inoperative. Further, investigators found that required testing of the engine alarms was not being routinely performed.

Alarm testing was part of COSCO's responsibility for ensuring that *Bright Field* maintain her "In Class" and "Periodic Unattended Machinery Space" certifications. That required, among other obligations, to periodically notify DNV of breakdowns, major repairs, and of successful tests of automated controls. They were also required to develop testing methods and procedures for the automated controls and monitoring systems. But during the Coast Guard's and National Transportation Safety Board's accident investigations, DNV representatives testified that they had not been made aware of *Bright Field*'s engine reliability problems.

COSCO also violated DNV regulations by failing to inform them of the inoperative #1 Terasaki monitoring system. Ocean voyages and inland waterway navigation were permitted with only one monitoring system operational, but failures of redundant or secondary systems were still required to be reported. In addition, there was no evidence that COSCO actively supervised the testing of the ship's automation system as required by DNV. The National Transportation Safety Board reported that:

> *The crew was not required to use the automated propulsion control test procedures that were delivered with the vessel in 1988 or to periodically test the functional operability of the automated propulsion control system. The engineering plant on the Bright Field was*

equipped with a number of safety devices, and the chief electrician submitted a report concerning the status of these devices to the vessel's owners quarterly. But the report was based only on a survey of alarms and sensors. No one regularly performed operational testing and maintenance of safety control devices such as the oil pressure switch that was supposed to start the standby oil pump in case of a drop in main engine lubricating oil pressure.

The Coast Guard recorded that:

> *The testing of the automation system to maintain unattended machinery classification consists of maintenance and function testing conducted quarterly by the vessel's crew and the testing done with the DNV surveys. On the BRIGHT FIELD, it is the responsibility of the Electrician to conduct the function testing. He testified that when doing this testing, he did not follow a test procedure and did not know of a test procedure that was onboard the vessel. This was consistent with the testimony of the other engineers that they did not know of an automation test procedure. The Electrician followed a COSCO form entitled, 'E.O. PLANT AND SAFTY (sic) DEVICES TEST RECORD' which listed the alarms and equipment that were to be tested. He used a list of alarm set points from an instruction manual onboard the vessel to determine if the test was satisfactory.*

Less Than Full Disclosure

Upon meeting the master of the *Bright Field*, the pilot asked if the ship's engine and navigational aids were in good working order. The master answered that they were. After major repairs were performed on the engine, the master may have believed the engine was in good condition, but he did not divulge any additional information as to the engine's repair history.

Had the pilot been aware of the engine's reliability history he could have made the decision to take extra precautions. These precautions may have included ordering tugs to escort the ship, adjusting transit speeds and maneuvering areas, maintaining closer vigilance of engine control activities on the bridge, or even delaying the voyage until optimum engine reliability was assured. Incomplete knowledge of the engine's failure history prevented the pilot from making important decisions concerning taking the ship downriver.

Learning From Bright Field

The theme of this chapter was that maintenance teams play a vital role in guaranteeing the integrity of the safety envelope. By maintaining equipment within prescribed bands, maintenance technicians place reliable machines into the hands of operating teams—machines that do what the designers intended. Conversely, if maintenance is deficient, reliability is compromised and the likelihood of breaching the safety envelope increased.

As we conclude, now is a good time to review some of the critical lessons of *Bright Field* and consider whether your team is susceptible to the same errors.

The Philosophy of Maintenance

Maintenance is a philosophy and an attitude. The philosophy comes from the top and is indicated by oversight, direction, and coordination of resources. In *Bright Field*'s case, COSCO failed to exercise the oversight and leadership necessary to set high maintenance standards aboard the freighter. The serious, repeated, and continuing problems with the main engine and lubricating oil system—all of which were known to COSCO leaders—demanded prompt and thorough analysis and repair instead of patchwork maintenance solutions.

Maintenance Underlies Reliability

Components and systems which are not maintained (or incorrectly maintained) often don't produce reliable results. We can't expect equipment to perform as designed if it is not kept in proper working order. Nor can we expect the safety envelope to offer the protection it advertises if the assumptions upon which it is based—one of which is that the subject equipment is currently within design specification—are not true. The *Bright Field* event is a clear illustration of this. The main engine and lubricating oil system were routinely outside of design tolerances. A review of the machinery history clearly establishes that fact. It is, therefore, not a surprise that the equipment failed to function reliably on December 14.

"Run It 'Til It Breaks"

The "run it 'til it breaks" approach to maintenance may be cheaper in the short run, but it is often far more expensive in the long run. (The *Bright Field* event illustrates the cost of such a philosophy.) The result is equipment which does not function as intended and challenges the operating staff to keep the systems running. Consequently, operators are tempted—or forced—to take short cuts, violate protective features, and engage in undisciplined (and often unsafe) habits of control. We have stated repeatedly that operations and maintenance are mutually supportive teams. What one does ultimately affects the other.

Preventive Maintenance

When industrial teams don't properly perform preventive maintenance, equipment problems tend to compound, creating a much more expensive repair problem in the end. Aboard *Bright Field*, fuel injector timing and lube oil system filter cleaning—two important elements of preventive maintenance—were either ignored or incorrectly performed. Excessive carbon and particulate buildup in the lube oil resulted, lending to the final loss of propulsion. Preventive maintenance is supposed to limit unnecessary corrective maintenance. If you don't do it, it won't.

Calibration of Instruments, Alarms, and Controls

Instruments, alarms, and equipment controls have tolerances within which they must control in order for the equipment's safety envelope to be valid. Hence, safety-related instruments must be aligned and calibrated on prescribed periodicities to ensure that the information they provide (and the functions which they control) are valid and offer the protection that they were designed to provide. The actions of *Bright Field*'s engineering staff to arbitrarily reset alarm set points for the lube oil sump level and lube oil filter differential pressure served to seriously degrade the main engine lubrication protective scheme. Considering the importance of propulsion to ship's safety, such actions are egregious violations of operating and maintenance standards.

Functional Checks

Functional checks—operating tests of equipment and systems—are designed to check reliability on a frequent (specified by technical manual) basis. Failure to perform them removes an opportunity to identify equipment problems before they are out of specification or not functional. Alarm and control testing aboard *Bright Field* was not being conducted in accordance with equipment technical manual requirements or DNV policies. As a result, the pilot and master were robbed of information regarding main engine performance at a critical juncture in the voyage.

System Alignment

For systems to function as designed (a key element in the definition of "reliability") they must be aligned as the designer envisioned. It is a predicate of safe operation and maintenance that components and systems be aligned and operated as prescribed in design documentation (and their derivative operating and maintenance procedures). If they are not, that portion of the safety envelope may be invalid. Aboard *Bright Field*, the failure of the engineering staff to align the second lube oil pump to STANDBY removed part of the ship's protection and ultimately resulted in a loss of propulsion at a critical time.

Warning Signs Most accidents are preceded by warning signs of impending failure. The accident is usually a final manifestation of problems that have been growing for some time—the tip of the iceberg, so to speak. In *Bright Field*'s case, the regularity of engine problems warned of the unreliability of the vital propulsion system. Though the freighter's officers were aware of the problems, their parent organization, COSCO, did not provide the support necessary to properly correct them, even in the face of repeated failures.

The Function of Lubricants Lubricants have to be clean, within specification for the equipment they lubricate, and sufficient in quantity to accomplish their objective. For *Bright Field*, the lesson of lubricants is self-evident. More recently, the Alaska Airlines crash off the coast of Los Angeles a few years ago was caused by a galled elevator jackscrew because of an improper lubrication regimen.

Case Study References

USCG U.S. Coast Guard Marine Board of Investigation. Investigation into the Circumstances Surrounding the Loss of Propulsion on the Bulk Carrier Bright Field…, Report Number 16732/MC97000078, December 8, 1997.

US Congress Gilcrest, Wayne T. – Presiding, Committee on Transportation and Infrastructure, House of Representative One Hundred Fifth Congress First Session. Congressional Review of the Bright Field Incident. May 14, 1997

NTSB National Transportation Safety Board. Allision of the Liberian Freighter Bright Field with the Poydras Street Wharf, Riverwalk Marketplace, and New Orleans Hilton Hotel in New Orleans, Louisiana, December 14, 1996. Report Number NTSB/MAR-98/01, January 13, 1998

Chapter 5

In Accordance With Approved Procedures

Procedures are instructions for the operation and repair of equipment. [See **The Industrial Operator's Handbook**, Chapters 3 and 9.] They provide road maps with check points (steps) to assist in controlling, maintaining, and testing machinery. Though they do not substitute for excellent training, they assist in task performance (if well-written) by reminding members of the operating and maintenance teams of the important steps in task accomplishment.

Since procedures are usually derived from equipment design documentation and manufacturers' technical manuals, their use is critical for operators and maintenance technicians to control equipment within—and restore equipment to—design specifications. But when operators and maintenance technicians begin to ignore procedures, improper operating and maintenance habits develop, and equipment is far more likely to be incorrectly operated or not restored to design specifications when repaired.

As we review the events of the Gannon Station accident of 1999, we would like to remind you that these were committed people with decades of experience who were trying to do a good job. Yet, they fell into the traps of complacency, miscommunication, and assumptions just as you and I can. Maintenance and operating procedures exist to help us avoid those traps. And that's the theme of this case study—maintaining equipment *in accordance with approved procedures!*

Gannon Station Explosion, April 1999

On April 8, 1999, at 7:15 A.M., an explosion occurred in Unit 6 of Tampa Electric's Gannon Station, killing three workers, injuring dozens more, and causing extensive equipment and facility damage. The explosion was caused by the release of hydrogen gas from a generator unit when a maintenance technician mistakenly removed a generator access cover before hydrogen had been evacuated from the generator cooling system.

Immediate Cause

While conducting scheduled maintenance on a turbine-generator as part of a maintenance outage, a maintenance technician from an outage maintenance group prematurely removed a generator casing access cover releasing 3,500 cubic feet of pressurized hydrogen gas used to cool the generator. The gas oxygen-balanced and was ignited by an unidentified ignition source. The ensuing explosion shook buildings as far away as six miles causing nearly $2 million dollars in damage to the generating facility. Three maintenance workers died and fifty other plant staff personnel were injured, some with severe burns. In addition to the deaths, injuries, and plant damage, TECO suffered extensive financial losses in lost electrical generation and generation replacement costs.

Contributing Factors

As we have emphasized again and again, accidents are almost always the result of a collage of factors. In **The Industrial Operator's Handbook** [Chapter 1, Anatomy of an Accident], we learned that seldom is a single unfavorable factor sufficient to support the occurrence of an accident. Rather, many factors are necessary to create the house-of-cards or domino-chain environment on which accidents are based.

In the Gannon Station event, the fatal removal of the Unit 6 generator access cover was only the last in a series of mistakes. Other factors that contributed to the accident included:

1. Failure to follow established procedures
2. Ineffective hazardous energy isolation
3. Inadequate job briefing
4. Incomplete communication
5. Poor leadership and supervision, and
6. Faulty teamwork

Preparing for Outage

Plant Description

Located near Port Sutton, Florida, Gannon Station is Tampa Electric Company's (TECO) second largest power generating facility. The station houses six coal-fired units, supplying high pressure steam to six steam turbine-driven generators. Rated at 1270 megawatts, the station serves Tampa and Hillsborough Counties, providing one third of the area's power needs. Only TECO's Big Bend station produces more power.

Figure 5-1: Aftermath of the Gannon Station Explosion. [Photo by The Tampa Tribune]

Hydrogen-Cooled Generator

Unit 6 turbine-generator is comprised of both high- and low-pressure turbines attached to a shaft which spins the rotor of a 375 megawatt electrical generator. The generator is cooled by both water and hydrogen. The stator employs a water cooling system, while the rotor and rotor-stator air gap are cooled by a closed hydrogen system, pressurized to approximately 45 psig.

Hydrogen is a volatile, flammable gas. In concentrations of approximately 4% to75% (by volume) in an air atmosphere, it is explosive. Perhaps the most graphic demonstration of the flammability of hydrogen occurred on May 6, 1937, when the German zeppelin, Hindenburg, exploded while docking at the Lakehurst Naval Air Station in New Jersey. Over 7.6 million cubic feet of hydrogen gas ignited, producing a fireball seen as far away as 15 miles.

Large electric generators like the ones used at Gannon Station develop considerable heat from both mechanical friction and the creation of electrical energy. Power production efficiency is inversely proportional to the process temperature. In other words, the higher the operating temperature, the lower the efficiency of the generator. Cooling is accomplished by circulating hydrogen gas through the generator in a closed loop system through heat exchangers.

Although highly explosive when mixed with air, the physical properties of hydrogen gas make it an ideal coolant for electrical generators. Hydrogen, having an atomic weight of 1, is the lightest known element, a characteristic that allows it to move through lines very quickly. For the same reason, hydrogen gas presents low resistance to the spinning generator rotor, reducing the heat of friction and wind effect (windage). Another advantage of hydrogen is its high rate of heat conduction. Finally, hydrogen is relatively inexpensive compared to more exotic gases.

Because of the potential for explosion and fire, Tampa Electric (and all of the power industry) has instituted stringent safety precautions for containing and controlling hydrogen in accordance with OSHA guidelines and manufacturer's recommendations. Procedural controls require that hydrogen be purged from the generator cooling system (and the cooling system be verified hydrogen-free) before commencing work that would necessitate breaching the generator casing.

Early Shutdown

The removal of a major power generator from the grid for outage maintenance is obviously an event that requires careful planning, preparation, and timing. Outages are timed when possible to occur during the yearly ebb tides of peak demand—times when air conditioning and heating demands are at their low points. Even then, power must often be purchased at premiums on the spot market to supply customer demand. Clearly, the longer a generating station is down for maintenance, the more generating revenue is lost, both in generation time and in spot market purchases. Yet, without excellent equipment maintenance, future reliability and safety could be jeopardized. Therefore, power plants (and most other industries) live by three strategic variables—safety, reliability, and cost. Next to proper operation, excellent maintenance is central to achieving all three objectives.

Gannon Station Unit 6 was scheduled to begin its annual maintenance outage on April 5, 1999. Anticipated work included inspection and repair of the boiler, turbine, and generator—most of which was to be accomplished by the company's own Energy Supply Group (ESG), a team of TECO maintenance technicians assembled as a traveling maintenance group twice a year to perform outage turbine and generator repairs. Scheduled ESG maintenance for Unit 6 generator included field removal, complete electrical inspection, testing of the windings, and a generator air test. Palm River regional power control station expected the Gannon Station Unit 6 to be returned to service by May 23, 1999.

Unit 6 was scheduled to go off-line at midnight on April 4, 1999, with outage work to commence shortly after the boiler, turbine, and other components were cool. However, approximately one week before the scheduled power outage, the boiler for Unit 6 developed serious problems requiring immediate attention. Unable to operate the boiler for another week, TECO decided to shut down Unit 6 earlier than originally planned and begin boiler repairs.

Energy Supply Group

Despite the early shutdown of Unit 6, turbine and generator repairs could not begin in conjunction with boiler work. The Energy Supply Group was still engaged in outage maintenance at TECO's Big Bend Station. They had to finish work at Big Bend before proceeding to Gannon Station. To expedite the transition of the ESG, Gannon Station leaders decided to perform as much preparatory work on the turbine deck as possible employing on-site maintenance personnel. A senior Gannon Station maintenance worker (a member of the ESG team) was assigned as temporary supervisor of the on-site work group. The team's task was to prepare the turbine-generator unit for disassembly by the soon-to-arrive ESG through staging of equipment, area cleanup, and removal of decking and other interferences. Specifically, the generator "dog house"—the generator exciter housing—was disassembled and removed.

Members of the Energy Supply Group began arriving at Gannon Station on Monday, April 5. With some of the repair work at Big Bend Station not yet completed, the ESG was fragmented between the two stations. Team members trickled in over the next few days.

The ESG was organized into two shifts—a day shift and an afternoon shift. Normally, each shift was led by a maintenance supervisor. However, the assigned supervisor for the afternoon shift was not available. In his place, a "step-up" journeyman (a senior maintenance technician on the team) was assigned as a substitute.

Commencement of Outage Work

Between April 5 and April 7, work on the turbine-generator unit proceeded according to the original project timeline. The premature shutdown had provided extra cool-down time of the turbine before its scheduled maintenance and allowed for maintenance personnel to get a head start on tasks scheduled later in the project.

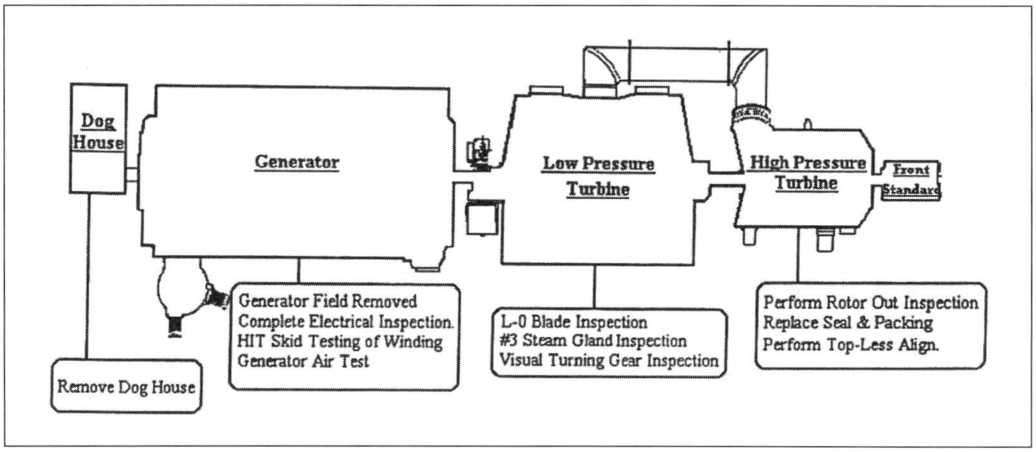

Figure 5-2: Generator and Turbine Arrangement. [From OSHA Report]

The boiler repairs scheduled in the first week were successfully accomplished even though additional work was necessary. Safety valves were removed for testing and adjustment. Slag was removed and appropriate surfaces were sand blasted clean.

On Tuesday, April 6, the generator pedestal cover ["Dog House", Figure 5-2] was removed. Later on Tuesday the stator cooling system was "burped". As much coolant as possible was drained by gravity through open valves in the stator housing. The remaining coolant was removed by pressurizing the system with air, then blowing it out.

Starting on April 7, electricians began generator stator testing using a special test unit called a "megger". Meggering is the process of electrically measuring stator and rotor winding insulation resistance. It is part of the process necessary to check for grounds within the stator and rotor. For reliable results, the test needs to be performed with the stator water cooling system drained, dry, and filled with hydrogen. After the megger, carbon dioxide is used to purge the hydrogen cooling system in preparation for opening the casing for inspection and repair. Pressurized carbon dioxide forces hydrogen from the system before the system is vented to atmosphere and filled with air. Failure to purge hydrogen from the generator cooling system before breaching the casing will release pressurized hydrogen to atmosphere, creating an extreme explosion hazard.

Meggering was scheduled for completion by midnight, April 7, at which time the hydrogen in the generator cooling system could be purged. However, testing did not go as planned. The electricians were plagued with problematic readings, presumably caused by a malfunctioning test unit. Electrical testing extended into the next day pushing the generator repair schedule back by at least 8 hours.

The Accident

Conditions on the Morning of April 8th

By Thursday morning, April 8 (the morning of the explosion), the top half of the high pressure turbine shell had been removed and the ESG maintenance technicians were ready to disassemble turbine blade rings. Some interference removal, cleanup, and organization for repair work remained to be done on the generator end, but the generator had not yet been purged of hydrogen since the plant electrical team had not yet successfully meggered the stator windings.

ESG Work Briefing

The day shift ESG supervisor arrived on site around 6:00 A.M.. He had received information from the electrical maintenance supervisor the previous evening that the electricians had "a good megger" and that the generator was "released" to the ESG. It is unclear with whom the ESG supervisor talked upon arrival at the site on Thursday morning, but he testified after the accident that he was aware that hydrogen had not yet been purged from the generator cooling system.

That morning, the remaining four members of the Energy Supply Group arrived from Big Bend Station, completing the day shift team. Prior to commencing work, the ESG day shift supervisor held a 10 minute morning work assignment meeting with his crew during which he told them that the stator cooling system was "dry", the electrical team had satisfactorily meggered the generator stator (a condition which had, in fact, not yet been achieved), and that the generator had been "released" to them. He did not mention, however, the state of the hydrogen cooling system or that the hydrogen system had not yet been isolated and tagged (since it had not yet been purged).

Missing from the ESG morning meeting was the operating crew watch engineer, a plant engineer assigned to the operating crew whose function, among many others, was to ensure warning tags were hung at hazardous energy isolation points before affected equipment was released for maintenance. Normally, the watch engineer attended the ESG work briefing and informed them of the status of the plant and work in progress. On this morning, he had been delayed at another meeting. He was one of several people who knew the status of the hydrogen cooling system in Unit 6. In fact, the warning tags for isolating the hydrogen system following purge and hydrogen-free verification were on his desk awaiting completion of the megger and a subsequent request by the ESG supervisor to purge hydrogen from the Unit 6 generator cooling system.

As the short morning meeting closed, work objectives were assigned to ESG members and the team reported to the turbine deck. Three maintenance mechanics were assigned to work the generator end of the unit, cleaning, removing remaining interferences, and staging parts containers for storing fasteners and components as they were stripped from the generator. Two were senior journeyman mechanics and the other was an apprentice who had volunteered to work the outage as a part of the ESG to accumulate the requisite training and experience necessary for permanent assignment to the ESG.

Confused Equipment Status

As ESG members left their morning work meeting, at least one senior journeyman mechanic—the mechanic that acted as step-up journeyman for the skeletal crew the week before—believed that hydrogen had been purged from the system. Yet, the ESG supervisor knew that the cooling system had not yet been purged and he believed that all of his team members understood the generator status.

Nor was confusion regarding equipment status confined to the ESG or to the state of the hydrogen cooling system. A few minutes before the explosion, a member of the plant electrical team came to the turbine deck to inform the ESG that meggering would soon commence from one deck below. His purpose was to warn the mechanics working near the generator about the potential electrical shock hazard posed by the meggering process. He told them that he would announce the start of the megger test over the public address system.

This information conflicted with what the mechanics had already been told by their ESG supervisor—that meggering was complete. Moreover, after the explosion, the electrician testified that he believed the generator cooling system to be hydrogen-free, though he knew it to contain hydrogen earlier in the week when the electrical team began the meggering process. Intra- and inter-team communications seem clearly to have been confused.

The Explosion

At approximately 7:15 A.M., an explosion occurred on the generator end of Unit 6 when one of the ESG maintenance mechanics assigned to prepare the generator for repair removed a casing access cover before hydrogen had been purged from the system. The ensuing explosion killed two of the mechanics—one journeyman and one apprentice—and another worker died from injuries resulting from the collapse of a portion of the outside of the generator building. Damage to the generator and other equipment housed in the building was extensive and approximately 50 people were injured.

The Unit 6 explosion caused TECO to shut down all units at Gannon Station until the cause of the explosion had been assessed. With Big Bend Station still off-line, the utility was forced to buy 800 megawatts of capacity on the spot market at a rate of between $50 and $100 per megawatt hour until late afternoon.

As a result of the explosion, the Unit 6 outage was extended by 6 weeks with additional repair costs of nearly $2.8 million. Gannon Station capacity was cut by about one-third for the duration of repairs.

Ironically, at the time of the explosion, the ESG supervisor was on his way to the control room to request that the hydrogen cooling system be purged and isolated.

Causes and Contributing Factors

Simply stated, the Gannon Station event occurred because the plant staff failed to control the configuration of equipment in a manner that would reasonably preclude releasing hydrogen to atmosphere from a pressurized generator cooling system. Instead, conditions were created in which one or more members of the Energy Supply Group believed the hydrogen to have already been purged from the system, leading to the fatal action of breaching hydrogen containment.

It is always tempting in accident analysis to conclude that an event was just the result of bad luck—the outcome of a few poorly-timed human errors. Yet, analysis almost always demonstrates that "accidents" are not accidents in the strictest sense at all. Rather, most occur because of compounded mistakes arising from undisciplined habits.

Worse, most are found to have been *easily* preventable through the application of very simple guidelines and techniques. Unfortunately, the evidence frequently reveals that those habits and techniques either had not been taught to the operating and maintenance teams, or else they had not been routinely and rigorously enforced. The Gannon Station event is not an exception. It occurred because unfavorable and non-conservative maintenance habits were allowed to accumulate and remain unchecked.

In **The Industrial Operator's Handbook** [Chapter 1, Anatomy of an Accident], we wrote that accidents are usually the cumulative result of a chain of apparently minor deviations from known standards, practices, and procedures. Undisciplined operating and maintenance habits such as failure to monitor critical parameters, poor communication (including unperformed or incomplete pre-job briefings), failed independent verification, and inadequate hazardous energy isolation are a few of the most common mistakes identified in **The Industrial Operator's Handbook** [Chapter 2, Common Components of Accidents].

In the Gannon Station event, no single factor or person was solely at fault. Instead, the action of the ESG maintenance technician to prematurely remove the Unit 6 access cover from the generator casing simply initiated the event much like the fall of the first domino in a domino chain or the placement of the last card in the collapse of a house of cards.

Underlying Causes Determining the initiating event is an important part of accident analysis; but of much greater consequence is learning *how* the domino chain or house of cards was built in the first place. To find the underlying or *root* causes of the Gannon Station event, one must go beyond the action of the ESG mechanic who lifted the access cover and ask, "What conditions allowed his errant action and why did they exist?" The "why did it happen" question is what differentiates root cause analysis from a simple origin and cause analysis which so often stops with identification of the initiating event.

Contributing Factors In asking "why" for Gannon Station Unit 6, several factors present themselves as candidates for accident contributors:

1. Failure to Follow Established Procedure
2. Ineffective Hazardous Energy Isolation
3. Inadequate Job Briefing
4. Incomplete Communication
5. Poor Leadership and Supervision
6. Faulty Teamwork

As we review each one, ask yourself whether disciplined performance in the area under consideration would've been an effective barrier to accident occurrence.

Failure to Follow Procedure

In **The Industrial Operator's Handbook** [Chapter 9, Understanding and Using Procedures], we stated that operating and maintenance procedures are instructions for running, preserving, and repairing machinery components and systems. They provide step-by-step guidance for starting, operating, stopping, and maintaining equipment.

We also stated that procedures do not substitute for well-trained operators and maintenance technicians. Rather, they are developed to *remind* well-trained personnel of the actions and the sequence of actions necessary to accomplish certain tasks. And, since memories of even the best fail, procedure review and use for complex, infrequently performed, and highly consequential tasks apply to seasoned experts as well as less-experienced people. Remember—airline pilots use procedures for "routine" tasks such as takeoffs and landings. They don't leave it to their memories.

Note also that procedures don't apply only to operators. Maintenance procedures are equally important. Just as an automotive shop manual prescribes the steps necessary to repair vehicle components, most equipment technical manuals also provide step-by-step guidance for disassembly, repair, and reassembly of components.

Relying on Memory In the Gannon Station event, a generator access cover was removed before the initial conditions (hydrogen purged and the hydrogen supply isolated) had been established for accomplishing that step. The question arises whether the ESG was following a procedure for performing their work.

After the accident, OSHA investigators asked the ESG day shift supervisor whether there was a procedure for disassembling the turbine-generator unit. Here are a few of the pertinent interview questions and responses:

Q. I guess what I'm getting at, is there some type of a sequential plan for the disassembly of a generator?
A. A sequential plan? There can be. I mean—
Q. So you do it from memory just basically because you've done it before?
A. Well, we've got little guidelines that we've printed up during the years. It's nothing as far as, I guess, gospel or whatever. It (sic) just little things like we'll put little sheets in there; do this, do that, the other. But everybody, I'm going to tell you, does things different ways.
Q. Is there a document out there somewhere that has these steps outlined?
A. There are. That's what I said. In the past, we've had them out there.
Q. I guess the question is, did you use any of the stuff that's out there to plan your activity?
A. No sir. I basically know what needs to be done. I mean, I don't—and most of my people know the same thing. They know as much as me, if not more. I mean, that's the nature of the business, as long as they've worked on it and stuff. But yeah, there are guidelines, steps, whatever you want to call it, a lot of them out there.
Q. So that afternoon if everything went according to plan and all the purging and tagging was done, you would have released your crew to pretty much do as their experience dictated?
A. Yeah, as long as it was purged and tagged and I told them it was safe to work on, yes, sir.
Q. In other words, they would be free to take it off at any point after that?
A. After the purging and tagging and safe to work on, yes, sir.

What the Procedure Said

Clearly, the ESG team members were relying upon memory in their approach to turbine and generator disassembly and repair. Yet, the ESG supervisor believed that there was a written procedure (or at least written guidelines) for the work that he and his crew were performing. During the investigation, analysts learned that there was, in fact, a procedure for performing the disassembly and inspection of the Unit 6 turbine-generator unit. The procedure, "Gannon 6 Generator Inspection Procedures", specifies the actions (and sequence of actions) to be performed for "Turbine End Disassembly", "Generator End Disassembly", and "Field (Rotor) Removal". Every step in each section has a block for initialing the completed step, an indication that the procedure was designed to be performed in a controlled, step-by-step fashion. The action taken by the ESG maintenance mechanic to remove the generator access cover is the 13th step in the "Generator End Disassembly" section. It reads:

13.___Unbolt and remove the two access covers from the upper half end bell. [See Figure 5-3, Generator Access Cover.]

Appropriately, the procedure is predicated by a section entitled, "General Information", a section (as in most procedures) that specifies initial conditions which must be established before performing the steps. The first two steps of that section read:

1.___Generator Stator Cooling System will be shutdown, drained and "burped". Megger of the stator and rotor will be done with the generator dry and in a hydrogen atmosphere.

2.___Once the initial testing is complete, the generator must be tagged and clearance received from Palm River Operations. The generator must be purged to air and the CO2 and H2 spool pieces removed and tagged. All of the turbine oil pumps and hydrogen seal oil pumps must be tagged.

The initial conditions clearly state that the generator stator and rotor winding resistance tests must be complete and the hydrogen cooling system purged and isolated before disassembly work commences. In this event, the requisite initial conditions had not been established.

It is unclear, however, whether the ESG supervisor and his crew had even seen the disassembly procedure. It is dated March 25, 1999, just two weeks before the accident. If they had seen the procedure, it seems doubtful that they had trained on it or even talked about it.

Would a Procedure Have Helped?

At this point, one might raise the objection that the existence, availability, and use of a procedure would have had no effect upon the mistaken action of the maintenance mechanic who removed the generator access cover, presumably because he thought the system contained no hydrogen or he didn't know what he was doing. Both arguments fail, however, when one considers that procedural steps for consequential activities are supposed to be consulted before the steps are performed. A maintenance mechanic should not have been removing any piece of the generator without knowing what step in the procedure he was performing. If it was the apprentice who took this action (assuming he was unaware of the danger), one of the senior mechanics on station should have been closely overseeing his activities and ensuring the disassembly procedure was followed. Do you recall the guiding philosophy for on-the-job training from **The Industrial Operator's Handbook**? We stated that *all actions related to the operation and control of plant equipment must be observed and verified at the time of performance by a person who is qualified (or certified) to operate the equipment.* The same rules apply to maintenance in which a performing technician is under instruction.

We must conclude that, had the ESG members been using their written procedure, they would have noted that the initial conditions section of the procedure had not yet been initialed (and, therefore, probably not established). Further, if they had been consulting the procedure, they would have realized that meggering had to occur in a hydrogen atmosphere. Since the electrical team warned members of the ESG that meggering was about to commence, it is more likely that ESG members would've questioned whether the generator was still filled with hydrogen.

Not a Panacea

The availability and use of written procedures for performing tasks which are complex, infrequently performed, or highly consequential (if incorrectly completed) is one of the safeguards employed in expedient and safe work performance

Figure 5-3: Generator Access Cover (Removed, center of photograph). [From OSHA]

for industrial operations. That's why in **The Industrial Operator's Handbook** we identified valid procedures as one of the strategy elements [Chapter 3, Systematic Industrial Operations] in disciplined operations and also one of the vital operating skills [Chapter 9, Understanding and Using Procedures].

But procedures are not a panacea against failure. They are "soft" barriers, much like highway speed limits. If not understood and observed, they are of little use. On the other hand, they offer excellent protection for well-trained technicians who have been schooled in the disciplined use of procedures and are committed to using them.

Ineffective Lockout/Tagout

Another valuable technique for protecting personnel and equipment is the process of isolating hazardous energy before commencing maintenance activities. OSHA Regulation 29CFR1910.269(d) specifies OSHA requirements for hazardous energy control (lockout/tagout) procedures as they apply to electrical power generation, transmission, and distribution activities. The purpose of the law is to offer reasonable assurance that every person working on or near a hazardous energy source during a maintenance activity is protected through isolation of the hazardous energy source.

In the Gannon Station event, one prominent hazardous energy source was pressurized hydrogen in the generator cooling system. The danger of concern was the explosive atmosphere that could be created by escape of hydrogen due to a cooling system breach before hydrogen was purged. Procedure was clear that the system should be purged and isolated before breaching the containment boundary, and the ESG supervisor and workers were well aware of the hazard. Yet, some of the workers believed that the hazard *had* been isolated—that the system had been purged of hydrogen and the hydrogen source blocked. They assumed that they were protected.

No Individual Locks or Tags

One of the tenets of OSHA law on hazardous energy isolation is that, for crew work, each worker should be individually protected by lock or tag. Paragraph (d)(8)(ii)(D) states:

> *Each authorized employee shall affix a personal lockout or tagout device to the group lockout device, group lockbox, or comparable mechanism when he or she begins work and shall remove those devices when he or she stops working on the machine or equipment being serviced or maintained.*

In other words, during crew work, each worker's lock (or tag) must be affixed to the group lockout (or tagout) before commencing work, and each person's lock (or tag) must be cleared before equipment can be re-energized. In this way, it is far less likely that a hazardous energy source will be mistakenly un-isolated while a crew member is still in jeopardy. Workers are not required to accept another's assurance that they are protected—a reasonable expectation because their own lives are at stake.

But at Gannon Station, the operating and maintenance teams did not adhere to that prescription. In practice, a maintenance supervisor requested of the operating staff that the system or component subject to repair be isolated and tagged. When isolation and tagging were completed, the crew tagging supervisor would affix the maintenance supervisor's name to a master tag (on which individual system tags were appended). The maintenance supervisor was then responsible to verify isolation, whereupon, if satisfied, he would direct his team members to commence repair work. Though maintenance team members had the right to personally verify isolation, few apparently did. Rather, their habit was to accept their leader's assurance that the subject system was safe for work.

If, however, ESG members had been required to affix personal locks or tags to a group lock or tag as required by law, they would have known that the hydrogen containment barrier for the generator was probably not yet ready for work since they had not yet hung tags or locks.

OSHA Citation

For this deficiency, OSHA cited TECO with a serious violation of hazardous energy isolation regulations. The citation reads, in part:

> *29CFR1910.269(d)(8)(ii)(D):When servicing or maintenance was performed by a crew each authorized employee did not affix a personal lockout or tagout device to the group lockout device, group lockbox, or comparable mechanism when he or she [began] work...For the Tampa Electric Company, Gannon Station Facility, Unit #6, authorized employees that were working on the morning of April 8, 1999, as part of the Energy Supply Maintenance Crew at Unit #6, were working under procedures that did not allow each authorized employee to affix a personal lockout or tagout device to the group lockout device, or comparable mechanism, when they begin their work.*

Gannon Station was also cited for three other serious lockout/tagout-related violations—1) failure to have an adequate lockout/tagout procedure, 2) failure to have an energy isolation verification process, and 3) failure to conduct periodic inspection of the energy isolation process.

We must conclude that, had federal regulations regarding lockout/tagout been followed, the explosion from an inadvertent release of hydrogen at Gannon Station would have been far less likely.

Not a Foolproof Protection

Note again that, as with procedures, the use of locks or tags for hazardous energy isolation is not a guarantee of safety. A tagging system especially can be rendered ineffective (as it was in this case) while the work group falsely assumes that it is protected. Locks offer more security, but both systems are dependent upon the thoughtful choice of isolation points followed by disciplined adherence to the prescribed system of protection. As we often admonished in **The Industrial Operator's Handbook**, there is no substitute for the alert, well-trained operator (or maintenance technician).

Protective systems are seldom foolproof. Rather, the scheme of protective barriers known as the safety envelope work together (much like the combined strength of the cords of a net) to reasonably reduce the risk of failure through a process known as *defense in depth*.

Inadequate Job Briefing

One of the most important cords in the safety net is the awareness of work team members regarding the status of equipment and related hazards. Operators and maintenance technicians usually leave their jobs and work teams after a period of duty, resuming their stations many hours (or sometimes many days) later. Usually, the job conditions and the equipment status have changed significantly in the interim.

Not surprisingly, deficiencies in knowledge regarding status and conditions (because of the time gap) frequently contribute to accidents. As a result, professional operators and maintenance technicians have established several helpful techniques to refresh their knowledge and awareness of job conditions prior to resuming work. Included in those techniques are proper job turnovers (through equipment walk-down, verbal information exchange with off-going workers, review of narrative logs and data record sheets), shift turnover meetings, and well-performed job briefings before commencing work.

The absence (or poor performance) of at least one of those techniques was a significant contributor to the Gannon Station event. Evidence indicates that the job briefing conducted in advance of the commencement of work on the morning of April 8 was inadequately performed.

In **The Industrial Operator's Handbook** [Chapter 8, Conducting Pre-Task Briefings], we defined a pre-task briefing as a preparatory meeting conducted before performing an operating, maintenance, or testing task. It is designed to ensure the safe and efficient execution of the task and usually consists of a review of the task steps, the role of team members in task accomplishment, a discussion of hazards characteristic of the task, and an examination of the safety constraints designed to protect against the hazards.

Job Briefing Content

Conspicuously absent in the job briefing described by the ESG supervisor for his work team on the morning of April 8 was a thorough discussion of the status of the generator unit and, in particular, the fact that the generator cooling system was still charged with hydrogen. Rather, the ESG supervisor simply told his people "the thing [the generator] has been released to us". And "it is dry". Apparently, several ESG members left the meeting with the impression that generator purging had already been accomplished and the hydrogen system had been isolated.

In a series of questions posed by OSHA investigators, the ESG supervisor described what he assumed his personnel knew:

> *Q. The picture here, these other two people up here, in your opinion, should they have been aware of the hydrogen state in the generator?*
> *A. Yes, sir. Because I never released it. And I specifically said—they've worked with me long enough. But I specifically said in the morning, all that was done was it was dry. That's all. The electric shop is done with it.*
>
> *Q. And they would know what that meant? I'm not trying to pin you down, but when you say it's just dry, they would know that that would mean that there's still probably hydrogen—*
> *A. Yes. Because I didn't tell them that it was purged.*
>
> *Q. Did [the two senior mechanics assigned to the generator] know that the generator was still in a hydrogen state?*
> *A. Well, he should have because he worked the unit the following week—previous week. He was down there leading the job. And he took the dog house off.*

OSHA Job Briefing Requirements

OSHA Regulation 29CFR1910.269(c) entitled "Job Briefing" requires that:

> *The employer shall ensure that the employee in charge conducts a job briefing with the employees involved before they start each job. The briefing shall cover at least the following subjects: hazards associated with the job, work procedures involved, special precautions, energy source controls, and personal protective equipment requirements.*

Note that OSHA law puts special emphasis on the need to identify hazards and the protections associated with those hazards during the briefing.

Obviously, though, not all jobs require job briefings. As we have seen for activities such as procedure use and independent verification, the ruling criteria revolve around complexity of the task, frequency of performance, and the potential consequences if the task is incorrectly performed. So, the question arises whether this job even required a job briefing—whether the ESG members should have reasonably understood the hazards and the status of equipment. A paragraph subsequent to the one cited above is entitled "Extent of Briefing". It states that:

> *A brief discussion is satisfactory if the work involved is routine and if the employee, by virtue of training and experience, can reasonably be expected to recognize and avoid the hazards involved in the job. A more extensive discussion shall be conducted if...the work is complicated or particularly hazardous, or...if the employee cannot be expected to recognize and avoid the hazards involved in the job.*

It is difficult to argue that disassembly of a turbine-generator set is routine, especially for a team which is constituted only twice a year for outages. Further, the presence of hydrogen in the generator cooling system made the job very hazardous. A mistaken action had the potential (and in this case actually resulted in) a fatal explosion. A thorough job briefing seems clearly warranted.

Moreover, a well-conducted job briefing wouldn't have required much more time than was expended. The ESG supervisor stated that the briefing lasted about 10 minutes—certainly not more than 15 minutes. Unfortunately, part of the briefing time was used to play a practical joke on the four ESG members who had just arrived from Big Bend Station.

OSHA Citation

As a result of the deficient job briefing conducted by the ESG supervisor on the morning of April 8, OSHA cited TECO for a serious violation of job briefing regulations. The citation reads:

> *29 CFR1910.269 (c): The briefing conducted by the person in charge did not cover the hazards associated with the job, work procedures involved, and special precautions associated with the work to be performed. For the work being conducted at the Tampa Electric Company, Gannon Station, Unit #6 Generator, the briefing that was conducted on the morning of April 8, 1999, did not effectively cover, or inform members of the maintenance crew, of the following hazards associated with the job, or special precautions and work procedures associated with the job:*
>
> *1) During the briefing the crew was advised that the electricians [were] getting a good megger and that the generator had been released to them, when in fact the electricians were still in the process of meggering the generator and were in the process of preparing to conduct another meggering test at the time of the explosion. The briefing did not advise the maintenance crew that the electricians were conducting the meggering tests with the generator containing a Hydrogen atmosphere.*
>
> *2) The Outage Schedule for Unit #6 called for the generator to be purged at 00:00 hours on April 8, 1999. The purging operation was not conducted as scheduled, and the morning briefing on April 8th did not inform the maintenance crew of the failure to purge the generator by the scheduled time and date.*
>
> *3) Unit #6 was 13 (thirteen) days into the scheduled outage, at the time of the explosion, and the Hydrogen had not been purged from the generator. Normally the Hydrogen is purged from the generator following tagging and clearance from Palm River Operations, or about 2 or 3 days into the outage. The morning briefing on April 8th did not inform the crew, or the experienced maintenance mechanics which just arrived at the Gannon Facility on the morning of April 8th, that the Hydrogen was in the generator for an extended period of time, or that the purging of the generator had failed to be performed by the date and time indicated on the Outage Schedule.*

4) On April 8, 1999, four experienced maintenance mechanics joined the crew that was already working at the Gannon Unit #6. Upon their arrival at their work locations it was obvious that the Turbines and the Generator were in various stages of disassembly. In particular the disassembly and removal of the Doghouse at the North end of the Generator indicated to the experienced mechanics that the outage was well under way and that they could continue the dismantling of the equipment that they came there to work on. The Gannon #6 Generator Disassembly/Inspection Procedure indicates that the removal of the Doghouse is normally done after the Generator is purged of the Hydrogen and the Turbine oil pumps and Hydrogen seal oil pumps are tagged out. The April 8th morning briefing did not inform the experienced mechanics, or any of the other crew members, that there were deviations in the normal Generator Disassembly/Inspection Procedures, so nobody on the crew had any reason to suspect that Hydrogen was still in the Generator, or that any other special precautions were necessary.

An Accident Stopper?

Job briefings are a final opportunity to ensure that team members know what they're about to do and how they should be doing it. In the Gannon Station event, the job briefing was, perhaps, the best of all opportunities to stop this accident from occurring. A bit of care and discipline in briefing performance would have surfaced the question of the status of hydrogen in the generator cooling system. Further, had the ESG been using their procedure for disassembly, identification of which steps they were about to perform would have been helpful in determining hazards and job status.

A Broken Team

In this short case study, we have reviewed three prominent contributors to the Gannon Station event—failure to follow established procedure, ineffective hazardous energy isolation, and inadequate job briefing. But questions regarding how the operating and maintenance staffs communicated, how they were led, and how they coordinated as teams remain.

Incomplete Communication

As we noted earlier, the confusion about the status of hydrogen in the generator cooling system was not confined to the ESG. The electrician who warned the team that meggering was about to recommence also believed that the system was hydrogen-free. We wonder whether outage team meetings (including the daily outage meeting) were being used to properly communicate the status of work in progress (and associated hazards) and coordinate the efforts of the various plant work groups. In well-disciplined plants, leaders of the operating team and their maintenance counterparts meet frequently to verify the status of equipment and the progress of work. It is a closely coupled process. At Gannon Station, the communication and coordination between the electrical maintenance group and the ESG, at least, was ineffective.

Poor Leadership and Supervision

Teams are only as effective as their leaders. Even teams with excellent players have difficulty winning consistently without good coaches. And most people who are selected for leadership roles want to do a good job. But many are ill-equipped for their challenges. Desire does not equal skill. In the Gannon Station event, we wonder how well prepared the ESG supervisors were for their roles. It is difficult to imagine the ESG team leader knowing the status of the generator cooling system while his players did not know. Some of his personnel were engaged in a task which he did not anticipate them performing. Nor did he believe that he had directed them to do so. This is a significant failure in communication. We must wonder whether the ESG supervisors were selected only on the basis of technical skills and knowledge (as is so often the case in U.S. industry), or whether they had also demonstrated acumen for leading people and coaching teams. Leadership is not a completely natural or intuitive skill. It requires training, practice, and mentoring.

Faulty Teamwork

The ESG was reconstituted twice a year for outages—a process that presents daunting challenges in the areas of leadership, training, coordination, and communication. They were tasked with moving from plant to plant to conduct turbine and generator work in a compressed time frame; yet, according to ESG members, they received no formal team training in their ESG configuration. Considering the importance of procedure compliance, hazardous energy isolation, and job briefings to the work management process, it seems that refresher training in at least those areas would have been appropriate each time an outage season began. It is very difficult for a team to be created (or, in this case, recreated) and then to commence their mission without some organization and training time.

Learning from Gannon Station

In the previous chapter, we focused on the theme that maintenance teams must maintain equipment within prescribed bands. In this chapter, we emphasized that well-devised maintenance procedures help alert, well-trained technicians to do exactly that. Even for team members with decades of experience, memories fail and communications can become confused as they did in this event. Procedures add order, control sequence, and serve as a baseline for performance—for maintenance teams as well as operating teams. As the National Transportation Safety Board reminded us, "...when applied in a disciplined, standardized manner [procedures] provide crewmembers with a firm foundation which they can depend upon for support during those times when they are subject to less than optimum levels of performance." [**The Industrial Operator's Handbook, 2nd Edition**, Chapter 9, "Understanding and Using Procedures".] As we review these final lessons of Gannon Station, consider whether your maintenance team has available and uses procedures for performing repairs in your plant.

Infrequent, Complex, High Consequence Tasks

No matter your level of experience, if you are preparing to perform an infrequent task that is complex in nature or high hazard, reference to a procedure is usually appropriate. At Gannon Station, the disassembly of Unit 6 generator was certainly infrequent and, with hydrogen not yet purged, very dangerous. The procedure for disassembling the generator established the need to purge hydrogen as an initial condition. It also provided in sequence the steps necessary for disassembly. Disciplined use of the available procedure would probably have prevented this tragic accident from occurring.

Hazardous Energy Isolation

Safety envelopes change as equipment conditions vary. In preparing equipment for maintenance (as was the case at Gannon Station), some of the usual protective barriers may need to be breached or degraded to accommodate repair. When protective barriers are removed, workers may be exposed to hazards that are not common to their normal work environment. In such cases, it is usually necessary to insert compensatory protective barriers (such as hazardous energy isolation measures) to ensure safety, especially for those performing repair work. In broader terms, as the dangers change, so must the protective scheme.

Where hazardous energy isolation is involved, operating teams must isolate components and systems before maintenance teams can add locks or tags in preparation for repair work. The controlling maintenance work document (which usually includes the procedure for performing the work) should include a step specifying that energy isolation be complete before potentially hazardous repair work may proceed. Further, the energy isolation lineup itself should be specified (again, as part of procedure) so that necessary isolation points are not forgotten.

At Gannon Station, the procedure for disassembly of the generator prescribed that hydrogen be purged and the source isolated before proceeding. Had it been used, the accident would have been far less likely.

Personal Protection

OSHA regulations (and common sense) require that maintenance team members individually lock (or tag) and personally verify their own safety from hazardous energy sources prior to commencement of work. This special form of independent verification provides an important barrier of protection to maintenance team members who are about to place life and limb in jeopardy. Unfortunately, the staff at Gannon Station did not enforce that tenet of hazardous energy protection. Instead, maintenance team members were in the habit of taking their leaders' word that protection had been established.

Pre-Job Briefings

Pre-job briefings are the final exam before commencing work. They provide one last opportunity to identify hazards, verify protection, coordinate the activities of team members, and plan for undesirable eventualities. At Gannon Station, the job briefing conducted by the ESG supervisor on the morning of the accident was ineffective at best. Though he knew that the hydrogen cooling system had not yet been purged, he failed to communicate system status to his team. Rather, he assumed that they were aware of status. Job briefings are, perhaps, the cheapest form of protection for a team. If you neglect them, you will subject your team to the same kinds of miscommunication errors experienced at Gannon Station.

Case Study References

OSHA

U.S. Department of Labor, Occupational Safety and Health Administration, Region Four, *Citations and Notification of Penalties* issued to Tampa Electric Company on September 7, 1999, for violation(s) of 29 CFR 1910.269 as determined by Inspection Number 109212571 and releasable supporting documents.

U.S. Department of Labor, Occupational Safety and Health Administration, Region Four, *OSHA Regional News Release*, USDOL: 99-197

U.S. Department of Labor, Occupational Safety and Health Administration, Region Four, *OSHA Regional News Release*, USDOL: 01-01

Department of Labor

Charles N. Jeffress, Dept. of Labor Assistant Secretary to Charles J. Kelly, Director, Industry Human Resources Issues Edison Electric Institute, *Clarification of Electrical Power Generation, Transmission, and Distribution*, November 21, 2000.

Local Investigation

Hillsborough County Fire Rescue FMO, Tampa, Florida, *Fire Incident Report - Incident 012081-00*, April 4, 1999.

Chapter 6

Last Line of Defense

The ultimate theme of this book has been that there is no substitute for alert, well-trained maintenance technicians maintaining equipment within specified bands in accordance with approved procedures. In pursuit of that theme, we have reviewed the concept of safety envelopes and the contribution of maintenance teams to envelope integrity. We have seen that, not only must the *hard* barriers—the physical constraints of the envelope—be in place, but also that the *soft* barriers—the mental boundaries of human understanding and performance—must stand in parity if the safety envelope is to remain intact. Team members must be not only well-trained, but also alert if the cords of the safety net are to protect. Furthermore, there is a need to restore equipment to design specifications to make the net functional. And, finally, it's important to perform the maintenance function in accordance with approved written guidance to help guarantee the effectiveness of the repair process.

Above all, however, the protections of the safety envelope are invalid if those who must control and maintain the equipment do not do so with vigilance and commitment. In that regard, we present one final case study for you to consider as you apply what you have learned to your own workplace.

Figure 6-1: Aftermath of the Amtrak Grade Crossing Collision. [From NTSB Report]

Railroad Grade Crossing Collision, September 1999

On September 26, 1999, a northbound Amtrak passenger train collided with an automobile westbound on U.S. Route 136 near McLean, IL, at about 5:08 P.M. (central daylight time). Both vehicle occupants—driver and passenger—were killed. The collision occurred where the Union Pacific Railroad's St. Louis Division main line and U.S. Route 136 cross near McLean, IL.

The Accident

Amtrak train 304-26 was traveling northbound en route from St. Louis to Chicago at about 74 mph on a "clear signal" (no track restrictions, speed limit 79 mph) as it approached U.S. Route 136. The automobile was traveling at an estimated 27 mph (speed limit 35 mph) westbound on U.S. Route 136 as it approached the grade crossing.

The Amtrak engineer was sounding the train's horn when he observed the automobile approaching the crossing from the east; the vehicle did not appear to be reacting to the horn. The engineer could not recall seeing the grade crossing warning devices operating, but he did recall that the car did not appear to weave around as typically seen when vehicles drive around lowered gates.

Evidence suggests that the occupants of the vehicle were unaware of the train's proximity until just before impact—the vehicle did not accelerate, take any evasive maneuvers, or leave any tire marks on the pavement that would have indicated an attempt to stop.

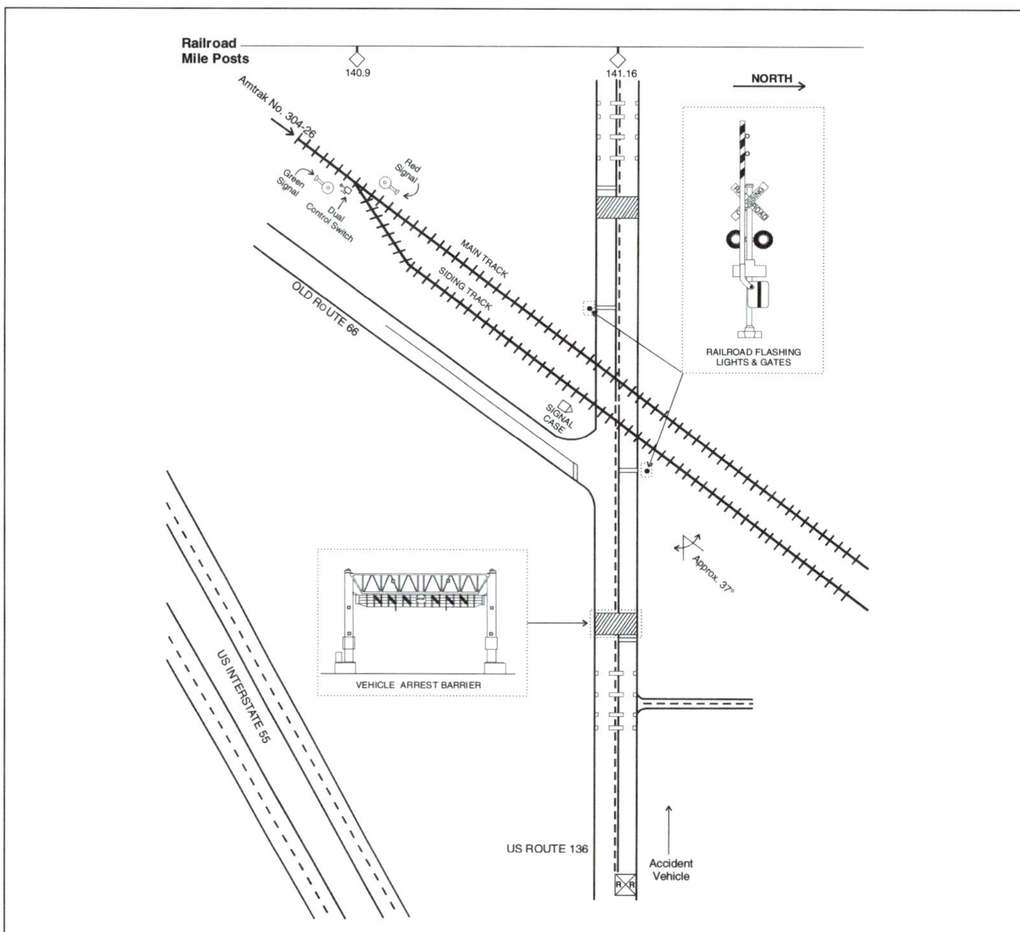

Figure 6-2: Accident Area Grade Crossing Layout Diagram. [From NTSB Report]

When the engineer realized a collision with the automobile was imminent, he placed the train's air brakes into emergency and braced himself inside the locomotive cab.

The impact split the vehicle into two parts. The front portion lodged on the front of the locomotive which came to a stop 3,216 feet from the crossing, while the rear portion came to rest about 203 feet north of the crossing and 7 feet east of the tracks. [See Figure 6-1.] Both automobile occupants were ejected from the vehicle and killed.

Immediate Cause The immediate cause of this accident was determined to be a signal maintainer's failure to remove a single jumper cable from the crossing control relay after performing maintenance and calibration at the crossing earlier that same day. That jumper cable prevented activation of both the warning lights and the crossing gates as the train approached. Hence, the vehicle driver was not provided with any active visual cues that it was unsafe to cross the railroad tracks.

As we will see in our later discussion, this "mistake"—forgetting to remove the jumper cable—was neither the only nor even the most significant factor related to this accident.

Description of the Grade Crossing

The U.S. Route 136 grade crossing [Figure 6-2] is an "active" crossing; the warning devices activate automatically through some form of train detection. This particular crossing was equipped with sophisticated, state-of-the-art detection, warning, and control systems.

The warning system is equipped with flashing lights, bells, and gate arms. Ten round (12-inch diameter) flashing light units mounted on two signal masts provide warning for all directions of highway traffic.

The grade crossing warning devices at this location uses a microprocessor-based system to calculate the speed of an approaching train and the time it will take for the train to reach the crossing. The microprocessor controls the relay logic (crossing control relay) that activates the flashing light units and raises or lowers the gate arms. The warning devices at the crossing are designed and calibrated to provide a minimum warning time of 21 seconds for all train speeds up to 79 mph.

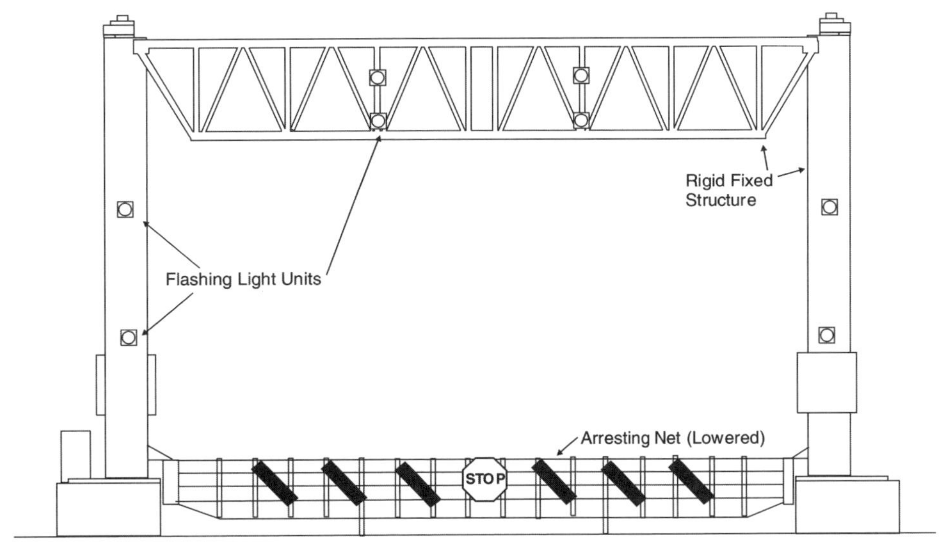

Figure 6-3: Vehicle Arresting Barrier Detail. [From NTSB Report]

The grade crossing predictor at the U.S. Route 136 crossing is also equipped with a data recorder and data recorder interface assembly which records information associated with train movements through that location. It records date and time of train movements, detected train speed, average train speed through the detection circuit, and the "island speed" (train speed at the roadway intersection). Additionally, it records 16 additional inputs monitoring the grade crossing predictor unit and the control system relay positions. This data recorder is quite similar to the "black box" data recorders on commercial aircraft.

This particular crossing was also equipped with an experimental vehicle arresting barrier system. [See Figure 6-3.] The system consisted of two barriers, one located on either side of the U.S. Route 136 crossing. Each barrier has a net assembly consisting of five horizontal wire cables suspended over the roadway between two towers connected by a fixed horizontal truss. The towers and connecting truss are equipped with eight 8-inch diameter flashing lights which activate when the net assembly begins to lower.

The vehicle arresting barrier is interconnected with the railroad warning devices such that a minimum 52 second warning time is given before arrival of a train.

Note, however, that this vehicle barrier system had been placed in "standby mode" since June of 1999 because of road work, system incidents, and system software updates, and was not in service at the time of the accident. Its relevance, however, is that the data event recorder and video monitoring equipment associated with the barrier system were operational and provided additional critical evidence about the accident.

Also note that the vehicle barrier study concluded in February, 2001, at which time all three such systems were taken out of service.

Pre-Accident Maintenance Events

The signal maintainer (age 55) was originally hired by a predecessor railroad of the Union Pacific in 1968, and had been working in the signal department for a total of 32 years. He had been responsible for the 21 grade crossings in the Bloomington territory which included the town of McLean for about 10 years.

Records indicate he was qualified as an employee-in-charge, flagman, lookout, lone worker, and machine operator; with additional qualifications in "signal test & maintenance procedures" and "on-track safety—all categories".

Callout At 12:30 P.M. on September 26, 1999, (the day of the accident) the UP Signal Operations Center called the signal maintainer for duty. The dispatch center reported problems with the south remote-controlled, power-operated switch on the McLean siding.

After arriving in McLean, the signal maintainer noticed that the grade crossing warning devices for the nearby Railroad Avenue crossing were operating (gates down, lights flashing) although there were no trains in the area. He determined that the grade crossing predictor unit had activated the warning devices due to a high and fluctuating "EZ" value (the operating voltage that monitors train traffic approaching the crossing). He corrected this problem by calibrating the unit, and then continued on to the McLean siding switch that he had been called to check.

Jumper Installed After completing his work on the power switch equipment (the originally assigned task), he decided to inspect the grade crossing predictor unit for the nearby U.S. Route 136 grade crossing where the accident subsequently occurred. He discovered that the "EZ" value for the siding was fluctuating—but not enough to activate the warning devices with no trains present. He then walked the siding track to inspect all track connections. He found and replaced a frayed "bond wire"—a connection to ensure electrical conductivity between track segments—and then returned to the signal box to recalibrate the grade crossing predictor unit.

In order to avoid stopping highway traffic while performing the calibration, he decided to use a jumper wire to falsely energize the crossing control relay [Figure 6-4] for the U.S. Route 136 grade crossing while he worked.

The crossing control relay is normally energized; when the system detects a train, it de-energizes the relay, which activates the warning lights and crossing gates. By artificially energizing the relay, the signal maintainer was temporarily circumventing the system's ability to activate the crossing's flashing lights and gates. (He later stated that he routinely used jumper wires in this manner to keep the warning devices from activating during maintenance work.)

He stated that after recalibrating the predictor unit, he "thought" he removed the jumper wire; and then notified the UP signal operations center that the south power switch (the originally reported problem he was assigned to repair) was back in service and operational. He did not inform the operations center that he had worked on any crossing equipment besides that south power switch. At about 4:30 P.M., he departed McLean en route to his residence.

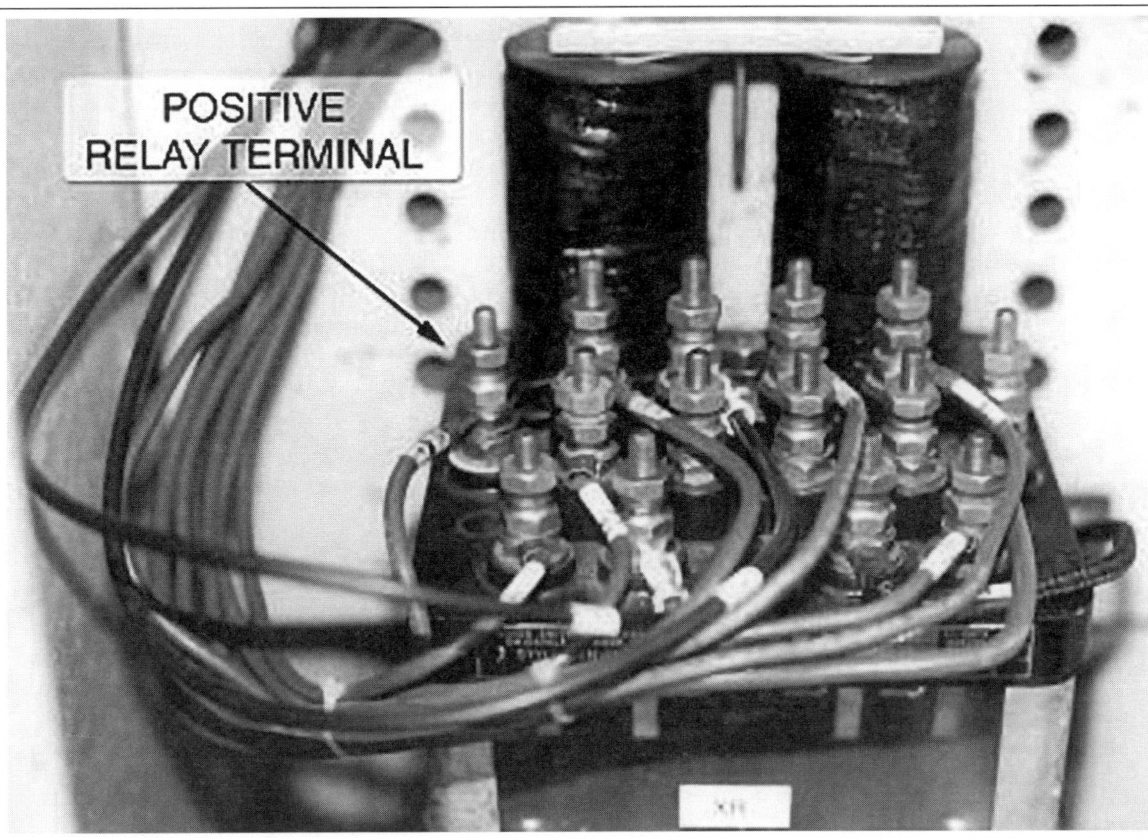

Figure 6-4: Crossing Control Relay. [From NTSB Report]

Maintenance Requirements Discussion

Before moving on to our discussion of the response and investigation, we will look at what "failed" with the work done by the signal maintainer.

Maintenance Requirements

Union Pacific (UP) signal tests and standards list all Federal Railroad Administration (FRA) required monthly, quarterly, and annual tests and inspections that must be performed on highway-rail grade crossing warning equipment. Additionally, the UP standards include general instructions for maintenance of warning systems.

UP procedures for adjusting grade crossing predictors state that:

> *A. Before Adjusting: Before changing the adjustment on crossing predictors or motion devices, the approach circuits in both directions must be walked, checking the rail, rail bonds, joint coupler termination shunts, correcting any defects.*
>
> *B. After Adjusting: When it is necessary to adjust crossing predictors or motion devices, shunt tests must be made on both approaches to check for the proper warning time for a train moving at maximum speed. A zero ohm shunt is to be placed at 90% of total approach distance and proper operation observed, or a train movement which activates the device accordingly must be observed.*

UP procedures also give the following directions about the use of test jumpers:

> *Test jumpers will be made of red test lead wire with red vinyl clip insulators. Jumpers will be at least five feet in length. Consideration should be given to limit the number of easily accessible jumpers to enable personnel to make a quick inventory before leaving the test location(s).*

The UP also defines precautions that must be taken when conducting tests or making changes. Requirements for protection for movement of trains include:

> *Maintenance changes or tests, which may interfere with safe operation of trains or other rail movements, must not be started until such movements have been fully protected. Temporary adjustments, when required, must be made in such a manner that the safety of trains or other operations will not be impaired. When adjustment, change, or replacement is made, tests must be performed immediately to determine that the apparatus functions as intended. When making tests of the apparatus, the proper instruments must be used and it must be known that no unsafe conditions are set up by the application of testing equipment.*

To comply with the requirements of providing protection for the movement of trains, the UP defines procedures in its maintenance-of-way rules. UP maintenance-of-way rule 137.2 states, in part:

> *When track, bridge, or signal personnel are working on the track in the approach to a crossing equipped with automatic warning devices, use one of the following methods to protect the public.*
>
> *137.2.1 Use one of the following methods to protect the public when working on a single main track.*
>
> *A. Positive Track Authority in Effect (track and time, track warrant, or track permit)*
>
> *1. No further protection is required because no train movements may be made over the crossing while the employee or gang holds that authority.*
>
> *B. Form B Track Bulletin in Effect*
>
> *1. The employee in charge must advise all trains during the clearance that:*
>
> *"The automatic crossing warning devices at MP [milepost] ___ are not working properly. You may proceed over the crossing at 15 MPH until the head end of your train completely occupies the crossing."*
>
> *C. No Positive Track Authority or Form B Track Bulletin in Effect*
>
> *1. Use a flagger(s) as follows to provide protection: One uniformed law enforcement officer, Two appropriately equipped flaggers.*

2. Request a Form C track bulletin that requires trains to proceed over the crossing at 15 MPH until the head end of their train completely occupies the crossing.
137.2.3 When making an emergency repair that causes the warning devices to activate and there are not enough employees to flag the crossing:
1. If no trains can operate over the crossing during the work, then no further protection is required.
2. If trains will operate over the crossing during the work, contact the train dispatcher directly and request that the dispatcher issue a 15 MPH slow order to all affected trains. This may require the dispatcher to contact trains by radio until either: The repair has been made– [or] A Form C track bulletin has been issued.

Deviations from Requirements

All of these methods require the maintainer to notify and/or obtain the necessary permissions/authorizations from additional railroad personnel before performing the maintenance. So, the signal maintainer deviated from UP rules in several significant ways.

First, he failed to obtain authority from the dispatcher while he worked on the track connection to provide protection against train movements for himself and for highway vehicles crossing the tracks while the crossing gates and lights were disabled by using the jumper to falsely energize the crossing control relay.

Second, he further deviated from the railroad's rules by not performing the required operational tests after recalibrating the grade crossing predictor. This was confirmed by data from the event recorders and video monitoring system.

If he had performed the operational tests, he would have immediately detected the failure of the warning devices to activate; and would have rectified the situation by removing the jumper wire from the crossing control relay.

Accident Response and Investigation

Following the collision, the Amtrak engineer used the radio keypad to initiate an emergency call-in, notifying the UP train dispatcher of the accident; the dispatcher informed the engineer that a signal maintainer would be dispatched to the scene to inspect the equipment. At about the same time, the Amtrak conductor placed an emergency 911 call using a cellular telephone. Emergency medical services personnel responded within minutes of the call; and a physician who was a passenger on the train offered his services at the scene. There were no reported injuries to the train crewmembers or passengers.

The UP signal operations center notified the UP manager of signal maintenance of the accident at about 5:36 P.M.; the manager directed the signal operations center to also contact and dispatch the area signal technician to the accident scene. (The signal maintainer was responsible for testing and maintaining all wayside and crossing equipment for his assigned territory, while the signal technician took care of the electronic components of this equipment for several territories; the manager of signal maintenance supervised both employees.)

The Illinois State Police had initial control of the accident scene and conducted the initial investigation; they assigned responsibility for accident scene reconstruction to the deputy coroner who also responded to the accident.

The signal maintainer and the signal technician both arrived (separately) at the scene at about 6:15 P.M.; the technician headed toward the crossing to assess damages, while the maintainer continued toward (and entered) the signal case which houses the crossing control equipment. After assessing damage at the crossing, the technician walked back and also entered the signal case.

The deputy coroner stated that, after arriving at the scene, he walked to the signal case after noticing that the door was open. Inside, he saw the signal maintainer (alone) holding some type of jumper wire while working on the equipment. The deputy coroner asked a State Police officer to secure the signal case; at about 6:25 P.M., the officer asked the two employees inside it (the technician had now also entered it) to exit the case and wait until the accident reconstructionist (deputy coroner) gave them authorization to begin testing.

During this period, a McLean police officer identified and interviewed five witnesses (bystanders near the crossing and a driver who had just crossed the tracks) who all stated that the crossing warning devices did not activate for Amtrak train 304-26. Upon hearing this, the signal technician notified the manager of signal maintenance of the situation and requested his presence at the scene.

The manager arrived about an hour later. With the manager, signal technician, and State Police present, data from the crossing event recorder was downloaded. The event recorder data indicated that the crossing control relay did not de-energize for Amtrak train 304-26; it also indicated that the track had been calibrated earlier in the day and that Amtrak 304-26 was the first train to travel through the crossing since the calibration.

The manager then asked the signal maintainer about previous work done at the crossing. The signal maintainer told the manager about the previously described replacement of the bond wire and his use of a red jumper wire to keep the crossing control relay energized during his recalibration of the grade crossing predictor unit; he further stated that he had finished his work at the crossing by removing the jumper wire.

Federal Railroad Administration (FRA) inspectors arrived at the crossing later that evening, taking control of the scene, and the subsequent investigation.

The crossing was tested numerous times on both September 26 and September 27, finding all relays and equipment to be properly functioning. On September 27, the crossing was completely retested with FRA inspectors present; those tests including re-enactments simulating train movements through the crossing—both with and without a jumper wire in place to keep the crossing control relay energized.

Following the re-enactments, the event recorder data was downloaded to compare with the data from the accident.

The vehicle arresting barrier system, although in "standby mode" as previously mentioned, was still functional with respect to its independent data recording and video monitoring systems. Video data from the six automatically actuated cameras indicated that:

- Amtrak 304-26 was detected as it approached the crossing (activating the monitoring equipment)
- The vehicle did not apply brakes when approaching the crossing (brake lights did not light)
- The grade crossing warning devices (flashing lights and crossing gate arms) did not activate with the approach of Amtrak 304-26
- The grade crossing warning devices did activate for the last train approach (before the accident and before the maintenance)

Data from both independent event recorders were in agreement (except for about a two minute difference in their clock settings) and indicated:

- Proper operation of all relays and warning systems for train approaches before the accident and the maintenance on September 26
- The maintenance performed by the signal maintainer was recorded on both data recorders
- Data on both recorders indicated that no testing of the crossing equipment was performed after the predictor unit recalibration

During the investigation, consideration was given as to whether an equipment malfunction might have caused the failure of the crossing warning system to activate. As just mentioned, analysis of the data from both event recorders and the video monitoring system showed agreement between all data recorded, and further indicated proper functioning of the crossing controls and warnings with train crossings both before and after the accident. Furthermore, complete and repeated testing was conducted on the crossing control relay after the accident. All tests indicated that the relay was operating within specifications and no deficiencies were identified that could have caused the relay to remain stuck in the energized position.

Questions of Integrity

First, the integrity of the accident scene evidence was compromised—perceptually, if not in fact—by the signal maintainer (who performed the maintenance at the crossing) being the first person to enter the signal box (unaccompanied) after the accident.

And, consider the additional evidence and statements regarding the "phantom" jumper wire:

- The deputy coroner saw the signal maintainer (alone) inside the signal box, holding some type of jumper wire while working on the equipment
- The signal maintainer told the signal technician (when he also entered the signal box) that he did not recall seeing or removing any jumper wires from the equipment when he (the maintainer) entered the signal box
- The signal maintainer told the manager of signal maintenance that he had indeed removed the jumper wire he used during the recalibration as the last step in finishing that work
- During accident investigation interviews, the signal maintainer told investigators that he "thought" he removed the jumper wire on the relay after recalibrating the predictor unit
- The manager of signal maintenance told the FRA that when he and UP attorneys met with the signal maintainer on September 27, the maintainer told them, "I must have left the jumpers on and killed two people".

Furthermore—besides the aforementioned questions of integrity—the signal maintainer was extremely "uncooperative" with accident investigators.

He refused to provide investigators with an account of his activities during the 72-hour period preceding the accident. Company records indicate he had been off duty for about 44.5 hours—since 4:00 P.M., September 24—before responding to the 12:30 P.M. trouble call on September 26, 1999.

He also refused to voluntarily undergo drug and alcohol testing after the accident—citing his "legal exemption" from such post-accident testing. Note, however, that he could have been required to undergo drug and alcohol testing "for cause" during the maximum 8-hour post-accident period—both because of his involvement in the probable cause of the accident and the fact that the event recorder data pointing to his involvement in the cause was available in an unusually timely period after the accident.

In light of this lack of cooperation, investigators were unable to determine whether or not fatigue or intoxication played any role in the cause of the accident.

Findings and Underlying Cause

Investigators reported the following findings.

> *1. Although the grade crossing warning system at U.S. Route 136 detected the approach of Amtrak train 304-26, the flashing lights and gates were prevented from activating because the Signal Maintainer had not removed the jumper wire that he had used to falsely energize the crossing control relay during the recalibration of the equipment.*
>
> *2. Because neither the crossing warning system nor the vehicle arresting barrier devices gave any indication of the approach of Amtrak train 304-26, the vehicle driver was not provided these active visual cues that it was unsafe to continue across the railroad tracks.*
>
> *3. Because the Signal Maintainer failed to fulfill the Union Pacific Railroad's requirement to conduct operational tests on the grade crossing warning devices after he performed a recalibration of the grade crossing predictor, he did not realize that the jumper wire was still attached to the crossing control relay.*
>
> *4. Had the Union Pacific Railroad provided its signal personnel with automatic warning equipment that alerted them when they failed to retrieve all jumper wires before leaving a work location, the Signal Maintainer would not have left the U.S. Route 136 grade crossing without removing the jumper wire that was still attached to the crossing control relay.*
>
> *5. Union Pacific Railroad's failure to secure the signal case immediately after the accident compromised the integrity of the accident investigation.*
>
> *6. The Union Pacific Railroad should have submitted the Signal Maintainer for reasonable cause toxicological testing but failed to do so.*
>
> *7. Exempting all railroad personnel from mandatory postaccident drug and alcohol testing following a grade crossing accident has the potential to exclude from testing obligations some railroad employees whose actions may have significantly contributed to the occurrence or severity of an accident.*

The NTSB determined that the probable cause of the accident was the failure of the signal maintainer to remove a jumper wire from the grade crossing control relay and, as required by the Union Pacific Railroad's written procedures, to verify the operational status of the grade crossing equipment after he had completed the maintenance work.

One final note about "personnel" actions taken after this accident: On September 27, the UP notified the signal maintainer that he was being removed from service for failing to follow UP procedures. Specifically, he was cited for failing to protect against train movements, violating roadway worker requirements (i.e., performing track maintenance without obtaining the necessary authorizations), and failing to perform required tests at the highway-rail grade crossing.

On March 10, 2000, the UP signal maintainer received a notice from the FRA for failing to comply with Federal regulations while working at McLean, Illinois, on September 26, 1999, citing the specific federal regulations he violated with respect to responsibility for ascertaining on-track safety, interfering with the normal operating functions of a rail system without providing for safety of highway traffic, and failing to test and determine proper functioning of the warning system before placing it back in service. Furthermore, the FRA determined that the signal maintainer was unfit for the performance of safety-sensitive functions and disqualified him from working in a safety-critical position on any railroad without direct supervision by a railroad manager or supervisor.

Learning from the Crossing Accident

Our purpose in this chapter has been to emphasize the role of individual technicians in securing the integrity of the safety envelope. In this accident, the crossing warning system was state-of-the-art. Yet, because a single individual chose to take a shortcut, the boundaries of the envelope were breached. If there is any lesson here, it is this: Barriers can't compensate for people who choose to bypass them! And nearly all protective barriers can be breached.

Individual Responsibility

The primary lesson from this occurrence is the same as for the first case we discussed in this book (Eastern Airlines maintenance failure to install the required O-rings on the L-1011 engine chip detectors). Once again, the last line of defense against disaster lies with reliance on each individual worker to perform assigned tasks accurately, professionally, and paying attention to the details. There is no substitute for alert, well-trained operators and maintenance technicians.

Furthermore, it doesn't matter how good their procedures are—and the procedures applicable to this accident were certainly accurate, comprehensive, and unambiguous. All it takes is one worker—the last line of defense—to ignore, circumvent, or otherwise fail in the performance of those procedures. In this case, that failure caused the death of two people.

Personal Integrity

And let's not forget one other—equally important—lesson to be learned from this accident: The importance of "personal integrity".

The signal maintainer's demonstrated lack of integrity—his attempts to "cover-up" his mistakes and uncooperativeness with the accident investigators—has implications that reach beyond this specific accident.

If integrity had been the rule—from a "moral" standpoint, if not a professional one—the true cause of the accident could have been determined within minutes or hours, rather than days and weeks.

The determination of accident cause was delayed by the signal maintainer compromising the integrity of the accident scene evidence, as well as by compromising the integrity of his own truthfulness in presenting a narrative description of his involvement in relevant causal events.

Fortunately, in this case, the supporting evidence obtained from the installed data recorders—combined with comprehensive, extensive, repetitive testing of the relay equipment—allowed the true cause of the accident to be determined with a degree of certainty.

Consider, however, what the outcome would have been if (as with many railroad signal boxes) such data event recording equipment was not installed. The actual cause of the accident might never have been determined. Furthermore, railroad personnel would be faced with a nagging uncertainty as to the reliability of the grade crossing warning relay equipment. They would be faced with the possibility that the relay system itself failed; and failed in a mode and for reasons which could be neither determined nor duplicated—and thus could be neither predicted nor prevented.

Trust, But Verify We cannot leave this sad chapter without identifying one additional lesson. Though the signal maintainer involved failed to conduct maintenance in accordance with procedure, the driver of the automobile struck by this Amtrak train apparently didn't exercise a fundamental tenet of defensive driving—look both ways before crossing a potentially dangerous intersection. Instead, he apparently assumed that, in the absence of a danger signal, there was no danger.

Case Study References

NTSB National Transportation Safety Board. Collision Of Amtrak Train 304-26 With A Highway Vehicle At A Highway-Rail Grade Crossing, McLean, Illinois, September 26, 1999. Report Number NTSB/RAR-0103, September 18, 2001.

Chapter 7

It All Depends on You!

As we began this book, we stated that safety in an industrial facility depends as much upon those who maintain equipment as it does upon the designers, builders, and operators. For it is through proper preventive and corrective maintenance that equipment is preserved and restored to design specifications—specifications on which safety envelopes are founded.

From that premise, we derived the theme of this text—that there is no substitute for alert, well-trained maintenance technicians maintaining equipment within specified bands in accordance with approved procedures. And, in support of that theme, we presented four case studies, each designed to illustrate a part.

Recounting the Lessons

From the Eastern Flight 855 near miss, we learned that training and alertness go together and that one does not guarantee (nor substitute for) the other. The mechanics involved were licensed and well-trained, and we have no reason to believe that they were not devoted to their work. Yet, their failure to pay attention to detail during a frequently performed "simple" task had far-ranging consequences. The chip detector O-rings formed redundant barriers on components (the detectors themselves) that penetrated a system (engine oil) vital to the survivability of the aircraft. As such, the O-rings formed a critical safety boundary. The procedures for replacing them were adequate, yet the mechanics did not ensure that the task was performed correctly. Had circumstances aligned a bit differently, instead of a near miss, this one could have been a hit that took the lives of 172 people.

In the *Bright Field* freighter accident, we saw how neglect to restore and preserve equipment (the main engine and lubricating oil system) to design specifications degraded the ship's propulsion system and auxiliaries to tolerances far below those prescribed by the designer and manufacturer. Aboard ship, of course, propulsion is vital to the protection of a vessel underway. But *Bright Field*'s protective margin for maneuvering was too thin for the circumstance encountered. The protection failed at the worst possible time. (Murphy always waits for best effect!) Moreover, though the timing of the failure was not predictable, the surety of it was. Repeated problems with the engine and lube oil system were a promise of future difficulties. As with Eastern 855, this one could have been much worse. Thankfully, the damage was limited to property destruction and a few evacuation injuries.

The Gannon staff was not so fortunate, though. Three people died to prove again the lesson that procedures are road maps for operating and maintenance tasks which remind us of the turns in the road and the hazards along the way for dangerous roads. Remember: It doesn't matter how well you think you know the route—when it's infrequently driven, complex, or filled with consequence when you take a wrong turn, then you're not very wise if you don't consult the road map. As we pointed out in **The Industrial Operator's Handbook, 2nd Edition**, if it's important for commercial pilots who fly every day, then it's important for you too when you're running your power plant, your oil refinery, or your chemical production facility.

Competence and Character

Finally, in the rail crossing accident, we observed how critical is the role of the individual technician in insuring the integrity of the envelope. The best training, the most thorough procedures, and the best equipment and processes are no match for people who, through intent or ignorance, fail to observe the practices that keep the margins of safety intact. A signal maintainer with over 30 years of experience repeatedly violated procedures, forgot to remove a jumper installed to perform a circuitry adjustment, failed to perform post work testing to verify the effectiveness of his repair, and then lied about his role in the accident. It seems that training cannot compensate for thoroughness, nor can experience compensate for lack of character.

It All Depends On You!

If there has been any message in this book and its parent (**The Industrial Operator's Handbook**), it is that equipment "accidents" are usually an outgrowth of undisciplined habits exercised by those who operate and maintain their machines and processes. But we have also found that almost all are preventable through the employment of the simple techniques that we described in Part III of **The Industrial Operator's Handbook**—habits that form the heart of the process we have described as systematic operations and maintenance.

Yet, those habits of disciplined operation and maintenance are not easily ingrained nor can your organization afford to enforce them through unending observation. Rather, their use depends upon your understanding of why they are necessary and your commitment to employ them *even when you think they are too conservative for the current situation.* For, just like seatbelts, you usually don't need them—but as surely as you relax the habit of using them, there will come a day that you regret it.

We hope that you have found your journey through **Maintenance and the Safety Envelope** to be both enlightening and practical. Thanks for letting us be your guide!